Pia Tölle

Proton Transport in Additives to Polymer Electrolyte Membranes

Pia Tölle

Proton Transport in Additives to Polymer Electrolyte Membranes

Material design for PEM Fuel Cells - An Approach by Computer Simulations

Südwestdeutscher Verlag für Hochschulschriften

Impressum/Imprint (nur für Deutschland/only for Germany)
Bibliografische Information der Deutschen Nationalbibliothek: Die Deutsche Nationalbibliothek verzeichnet diese Publikation in der Deutschen Nationalbibliografie; detaillierte bibliografische Daten sind im Internet über http://dnb.d-nb.de abrufbar.
Alle in diesem Buch genannten Marken und Produktnamen unterliegen warenzeichen-, marken- oder patentrechtlichem Schutz bzw. sind Warenzeichen oder eingetragene Warenzeichen der jeweiligen Inhaber. Die Wiedergabe von Marken, Produktnamen, Gebrauchsnamen, Handelsnamen, Warenbezeichnungen u.s.w. in diesem Werk berechtigt auch ohne besondere Kennzeichnung nicht zu der Annahme, dass solche Namen im Sinne der Warenzeichen- und Markenschutzgesetzgebung als frei zu betrachten wären und daher von jedermann benutzt werden dürften.

Verlag: Südwestdeutscher Verlag für Hochschulschriften GmbH & Co. KG
Dudweiler Landstr. 99, 66123 Saarbrücken, Deutschland
Telefon +49 681 37 20 271-1, Telefax +49 681 37 20 271-0
Email: info@svh-verlag.de

Zugl.: Bremen, Universität, Diss., 2011

Herstellung in Deutschland:
Schaltungsdienst Lange o.H.G., Berlin
Books on Demand GmbH, Norderstedt
Reha GmbH, Saarbrücken
Amazon Distribution GmbH, Leipzig
ISBN: 978-3-8381-2808-5

Imprint (only for USA, GB)
Bibliographic information published by the Deutsche Nationalbibliothek: The Deutsche Nationalbibliothek lists this publication in the Deutsche Nationalbibliografie; detailed bibliographic data are available in the Internet at http://dnb.d-nb.de.
Any brand names and product names mentioned in this book are subject to trademark, brand or patent protection and are trademarks or registered trademarks of their respective holders. The use of brand names, product names, common names, trade names, product descriptions etc. even without a particular marking in this works is in no way to be construed to mean that such names may be regarded as unrestricted in respect of trademark and brand protection legislation and could thus be used by anyone.

Publisher: Südwestdeutscher Verlag für Hochschulschriften GmbH & Co. KG
Dudweiler Landstr. 99, 66123 Saarbrücken, Germany
Phone +49 681 37 20 271-1, Fax +49 681 37 20 271-0
Email: info@svh-verlag.de

Printed in the U.S.A.
Printed in the U.K. by (see last page)
ISBN: 978-3-8381-2808-5

Copyright © 2011 by the author and Südwestdeutscher Verlag für Hochschulschriften GmbH & Co. KG and licensors
All rights reserved. Saarbrücken 2011

Acknowledgement
The presented work corresponds to the authors dortoral thesis in the group of **Prof. Thomas Frauenheim**, BCCMS at university of Bremen. The author would like to acknowledge the close collaboration and the advice by **Dr. Christof Köhler** and the proofreading by **Svea Sauer** and **Dr. Giulia Tomba**. Furthermore, the author would like to thank the Deutsche Forschungsgemeinschaft for the financial support in the course of the project SPP-1181.

Abstract

The enhancement of proton transport in polymer electrolyte membranes is an important issue for the development of fuel cell technology. The objective is a material providing proton transport at a temperature range of 350 K to 450 K independent from a purely water based mechanism. To enhance the PEM properties of standard polymer materials, a class of additives is studied by means of atomistic simulations consisting of functionalised mesoporous silicon dioxide particles. The functional molecules are imidazole or sulphonic acid, covalently bound to the surface via a carbon chain with a surface density of about 1.0 nm^{-2} groups.

At first, the proton transport mechanism is explored in a system of functional molecules in vacuum. The molecules are constrained by the terminal carbon groups according to the geometric arrangement in the porous silicon dioxide. The proton transport mechanism is characterised by structural properties obtained from classical molecular dynamics simulations and consists of the aggregation of two or more functional groups, a barrier free proton transport between these groups followed by the separation of the groups and formation of new aggregates due to fluctuations in the hydrogen bond network and movement of the carbon chain. For the different proton conducting groups, i.e. methyl imidazole, methyl sulphonic acid and water, the barrier free proton transport and the formation of protonated bimolecular complexes were addressed by potential energy calculations of the density functional based tight binding method (DFTB). For sulphonic acid even at a temperature of 450 K, relatively stable aggregates are formed, while most imidazole groups are isolated and the hydrogen bond fluctuations are high. However, high density of groups and elevated temperatures enhance the proton transport in both systems.

Besides the anchorage and the density of the groups, the influence of the chemical environment on the proton transport was studied. Therefore, the uptake and distribution of water molecules was estimated from classical molecular dynamic simulations and the local chemical environment was determined for different functional groups. The sulphonic acid functionalised silicon dioxide pores are more hydrophilic than the unfunctionalised and the imidazole functionalised systems. At lower hydration, the distribution of water is inhomogeneous and the surface of the pore is covered by a water layer for all systems. In addition to the interaction with water, an interaction of functional groups with the surface is observed which is shielded under hydration. Due to these interactions, the number of isolated groups and their stability is increased under the influence of the environment that reduces the proton transport mechanism which has been described before.

Apart from the proton transport mechanism known from the vacuum system, two additional mechanisms occur under the chemical environment. These mechanism directly involve water molecules. One possibility is the complete deprotonation of the functional group, followed by water based proton transport as expected for acidic system, e.g. sulphonic acid. Another possibility is a water based proton transport over short distances from one proton conducting group to another. The three competing mechanisms are studied by free energy calculations and their occurance is evaluated according to the local environment conditions. The proton transport mechanisms involving water are more favourable in sulphonic acid functionalised particles, while the dominating mechanism is comparable to the mechanism in vacuum for imidazole system.

Contents

Introduction   1

I   Basic Principles and Approaches to the Topic   5

1   Principles of Computer Simulations   7
- 1.1   Particle Interactions and Forces 7
  - 1.1.1   Classical Force Fields 8
  - 1.1.2   The Electronic Problem - Ab Initio 9
  - 1.1.3   The Electronic Problem - Density Functional based Tight Binding . . 13
  - 1.1.4   The QM/MM Approximation 17
- 1.2   Classical Molecular Dynamics Simulations 18
  - 1.2.1   Thermodynamic properties 18
  - 1.2.2   Different Thermodynamic Ensembles 20
- 1.3   Chemical Reactions 24
  - 1.3.1   Activation Energy 24
  - 1.3.2   Umbrella Sampling 24
  - 1.3.3   The Weighted Histogram Analysis Method 26
- 1.4   The Proton Coordinate 27

2   Proton Transport and PEM Additives   31
- 2.1   Protonation of Molecules and Proton Transport 31
  - 2.1.1   Proton Transport in Water 31
  - 2.1.2   Other Proton Transport Species 32
- 2.2   PEM Additives - Experimental Insight 34
  - 2.2.1   Characterisation of the Functionalised MCM-41 34
  - 2.2.2   Proton Transport in the Material 36

II   PEM Additives - Computer Simulation   39

3   Proton Transport Species   41
- 3.1   Proton Conducting Species - Validation 41
  - 3.1.1   Model Systems and QM Methods 41
  - 3.1.2   Binding Energy 42
  - 3.1.3   Proton Affinity 43
  - 3.1.4   Proton Transport Barrier 45
  - 3.1.5   Short Conclusion 47
- 3.2   Proton Transport Ability of Functional Groups 47

	3.2.1 Model System of Proton Conducting Groups in Vacuum	48
	3.2.2 Diffusion of an Excess Proton	48
	3.2.3 Structural Properties	52
	3.2.4 Short Conclusion	59

4 Functionalised Silicon Dioxide Material — 61
4.1 The Amorphous Silica Model and FF Simulations — 62
    4.1.1 Model System — 62
    4.1.2 Computational Details — 62
4.2 Fully Hydrated Environment — 64
    4.2.1 Average Water Density — 64
    4.2.2 Average Number of Water Molecules per Group — 66
4.3 Different Humidity — 66
    4.3.1 Density Profile — 66
    4.3.2 Local Water Environment — 68
    4.3.3 Interaction of Groups with the Surface — 70
4.4 Short Conclusion — 70

5 Proton Transport Inside the Porous Environment — 73
5.1 Free Energy Barrier Calculation and Reaction Coordinate — 73
5.2 Water Based Proton transport — 75
    5.2.1 Free Energy of Deprotonation — 76
    5.2.2 Vehicular Diffusion Coefficient in Sulphonic Acid System — 77
5.3 Proton Transport involving Functional Groups — 78
    5.3.1 Direct Transport between two Groups — 79
    5.3.2 FE of Water based Transport between two Groups — 81
    5.3.3 Direct Proton Transport on the Substrate[1] — 83
5.4 Short Conclusion — 89

Summary and Discussion — 91

A Additional Data — 93

B Changes in Gromacs 4.0.5 — 95
B.1 Usage of the dftb+ QM/MM — 95
B.2 Usage of the mCEC-coordinate — 96
B.3 Details of the Implementation — 97
    B.3.1 Derivatives of the mCEC — 97
    B.3.2 Derivatives of the Reaction Coordinates — 98

C Force Field Parameters — 99

Bibliography, Colophon and Lists of Symbols and Abbreviation — 111

Proton Conducting Membrane for Fuel Cell Application

Historically, the beginning of fuel cell technology can be traced back to the mid of the 19th century, when C.F. Schönbein and W. Grove built the first fuel cell with liquid sulphonic acid as the electrolyte membrane splitting a reaction in two half-cell reactions. It took more than one century before prototypes were built for applications still being exclusive machinery, such as for space programs and military purposes. Nowadays, fuel cells are considered to be a promising "green" technology with high power density and efficiency. The technology opens the perspective to decrease the dependence on the oil industry by exploitation of hydrogen. Additionally, it provides emission-free car engines, promising clean air despite increased traffic volume inside urban areas. The ongoing development of this socially supported research field is encouraging, but commercial applications develop slowly especially due to high costs and missing infrastructure for the fuel supply. So far, only pilot projects for the automobile industry were successfully launched in many countries.

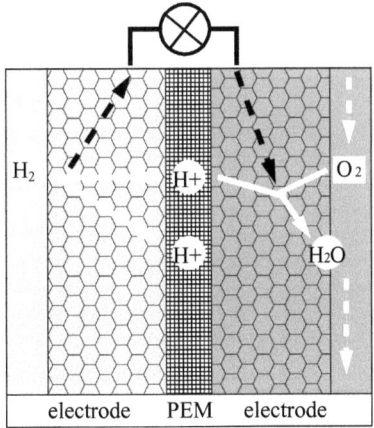

Figure 1: *Schematic view of a polymer electrolyte based fuel cell; white arrow - flow of reactant; black arrow - flow of electrons; The dashed lines mark exchange with the outside. PEM and electrodes are shown as different patterns; half-cell of oxidation reaction (light grey) and of reduction reaction (dark grey).*

Different classes of fuel cells can be distinguishes depending on its components or fuel gases, i.e. reactants. The membrane material influences significantly the performance and properties of the fuel cell system, e.g. the operating temperature. Important classes of membrane materials are alkaline or acidic electrolytes, solid oxides or polymer material which is referred to in the following. For these systems, primary goals for research are water and heat circulation as well as enhanced durability and reliability of materials[2, 3].

In Fig. 1, a schematic view of a polymer electrolyte based fuel cell is shown. A redox reaction is split into two half-cell reactions by the *polymer electrolyte membrane* (PEM) which avoids a direct contact between the reducer, e.g. hydrogen, and the oxidiser, e.g. oxygen. In the half-cell containing the reducer, protons and electrons are separated from the oxidised molecules. The electronic current to the other half-cell is exploited as the PEM is electronically isolating, while protons are conducted to the other half-cell and are reduced with oxygen to water. Today's standard PEM fuel cells consist of a 20 to 200 μm polymer joined at both sides with several hundreds μm thick electrodes of porous carbon fibres functionalised with Pt nanoparticles[2]. This complex system provides many challenges for computer simulation research[4, 5]. This work focuses on the PEM. In general, the PEM needs to accomplish the following demands: high proton conductivity (and intrinsic charge carrier density), electrical isolation and gas (both reactant, oxidiser and reducer) blocking ability. The material needs to be mechanically stable and no degradation should occur during service, i.e. the membrane needs high thermal stability and high chemical resistance to peroxide and hydro peroxide, which occur as side-products of the process.

The commonly used PEM Nafion® - developed by DuPont in the late 1960s - is a perfluorosulphonic acid membrane that has an excellent chemical and thermal stability, but it is quite expensive and difficult to recycle or dispose (being a perfluorinated material)[6]. Consisting of hydrophilic and hydrophobic groups, the nano-phase-separated structure provides a connected water network only under a certain hydration[7, 8]. In low humidity conditions, the hydrophilic regions are instead separated from each other by the fluoronated carbon backbone[8, 9, 10]. As proton transport in Nafion is mainly water based, the high proton conductivity of Nafion is only ensured under well defined thermodynamic conditions that provide sufficient hydration of the membrane, i.e. temperatures below about 100°C at ambient pressure [9]. Such a narrow temperature range represents a serious limitation to the fuel cell system, especially as the advantages at an elevated temperature are compelling. Higher temperature enhances the rates of electrochemical kinetics, which means higher efficiency. Costs are lowered as the water management and cooling is simplified and the CO tolerance of the electrodes (anode) is increased, which allows the use of cheaper impure hydrogen gas. The advantages are explained in detail in reference [3].

Many approaches have been proposed to improve the membrane properties at elevated temperature and low hydration. Three main strategies occur: firstly, the replacement of water in its function as proton solvent, secondly, the replacement of the inert polymer matrix by a matrix with intrinsic proton conductivity and thirdly, the incorporation of inorganic compounds[6]. The first approach has already shown promising results, e.g. for a Nafion membrane filled with liquid imidazole, the conductivity of the membrane was comparable to the water-swollen Nafion at a about 100 K higher temperature[6]. However, leakage of proton solvent is still a problem in such a material, augmented even by the risk of poisoning the platinum electrodes with imidazole. For the second approach, local mobility of the proton solvent is of high importance. Polymer materials containing hetero cycle groups like benzimidazole or imidazole and phosphorous acid bound to side chains and the promising sulphonated poly ether-ether ketone (S-PEEK) have been reported and high dependence of conductivity on chain length and flexibility are found[6]. Finally, mechanical strength, thermal stability and water retention at elevated temperatures are improved by the addition of inorganic compounds (SiO_2 or TiO_2) to the membrane[3, 11].

In an interdisciplinary project, computational approaches as well as experimental methods were applied to study and develop a hybrid systems of organically functionalised porous silicon dioxide particles. The aim of the novel material is an enhancement of PEM properties when adding particles to standard polymer material. The functional groups are imidazole or sulphonic acid covalently bond to the surface via a carbon chain to prevent leakage. The main interest of the computer based research lies on the description of the proton conducting abilities of the novel material. Since this project is in close collaboration with experimental research, the material and parameter ranges of the described computational study are chosen facing also experimental limitations. Besides hydrogen bond fluctuation, the mobility of proton conducting molecules strongly influences proton transport. Being covalently bound to the inorganic surface, the mobility of functional groups is not comparable to functional groups incorporated in a polymer material or to liquid or crystalline proton conductors. Therefore, the effect of immobilisation via carbon chain and local mobility as well as the influence of chemical environment on the dynamic of the immobilised groups and on the proton transport are described. The focus lies on low humidity conditions at a temperature of about 450 K corresponding to the critical conditions for the performance of PEM and on the dominant proton transport mechanisms in these chemical environments. The analysis of structural proton transport demands sufficient accuracy and a quantum mechanical description of reactions, whereas the chemical environment is influenced on large length and time scales. Therefore, different atomistic descriptions are needed, namely classical force field and quantum mechanical methods. The multiscale aspect of the present study is accomplished by the comparison of results from separated simulations on different length scales as well as by directly coupled quantum mechanical - molecular mechanical simulations.

In the first chapter, the fundamentals of the applied computational methods are described shortly. In the second chapter, proton transport of water and the functional groups is reviewed and the hybrid material is introduced as described by experiment. In the third to fifth chapter, the results from computer simulations are explained in details. Firstly, the proton transport species are analysed, the quantum mechanical method is validated and the effect of immobilisation on the proton transport is studied. Secondly, the silicon dioxide material is described and the chemical environment of proton conducting groups inside the material is characterised, especially regarding dry conditions as critical operating conditions of the fuel cell. Finally, different proton transport mechanisms are identified and evaluated for both functional molecules in the novel material with respect to different functionalisation.

Part I

Basic Principles and Approaches to the Topic

Chapter 1

Principles of Computer Simulations

In this chapter, theoretical methods that are used in the present work are introduced shortly, beginning with the description of particle interactions, Sec. 1.1. The atomic system are described applying the Born-Oppenheimer (adiabatic) approximation, which allows the decoupling of the electronic and nuclear motion, e.g. of the differential equation of the electrons and the nuclei, due to the difference in their masses. The atomic interactions are described either through an empirical *force field approach* (FF), see Sec. 1.1.1, or derived from a *quantum mechanical description of the electronic part* (QM), see Sec. 1.1.2 and 1.1.3. In Sec. 1.1.4, *quantum mechanical molecular mechanical coupling* (QM/MM) is discussed. Based on the atomic interactions, *atomistic molecular dynamics simulations* (MD) are performed under different thermodynamical conditions in order to obtain dynamic properties, see Sec. 1.2. In Sec. 1.3, basic ideas of transition state theory conduct to the evaluation of chemical reaction by energy barrier and umbrella sampling is introduced as a simulation method of free energy pathes depending on a reaction coordinate. The proton coordinate is defined in the last section, which enables the description of proton transport reaction. All formulae are given in atomic units[i].

1.1 Particle Interactions and Forces

In this section, different approaches to atomic interactions are explained. In subsection 1.1.1, classical force fields are introduced as empirical descriptions of atomistic interactions on the molecular scale neglecting details about the electronic structure. Therefore, this method of low computational cost is commonly used for MD simulations, discussed in Sec. 1.2. In subsection 1.1.2 and Sec. 1.1.3, QM approaches are described with particular focus onthe *Hartree-Fock method* (HF), on the *density functional theory* (DFT) and finally on the *density functional based tight binding method* (DFTB). These methods also provide atomic forces for MD simulations as the derivative of the energy expression with respect to the atomic positions and - in contrast to the classical FF method - cover chemical reactions and charge transfer. A QM/MM coupling scheme, as discussed briefly in section 1.1.4, combines the advantages of both types, FF and QM, enables simulations of long time and length scales while providing a detailed description of reactions and charge transfer in certain regions of the system.

---

[i]Constants are set to one: charge of an electron e=-1, Planck constant $\hbar$=1, vacuum permittivity ϵ=1, electronic mass m_e=1.

1.1.1 Classical Force Fields

A *classical force field description* (FF) consists of the empirical parametrisation of atomic interactions neglecting any electronic structure information and the energetic levels, e.i. the band structure. The interaction is divided into inter and intra molecular terms with a given functional relation and a small number of (free) parameters. Due to the empirical character of the potential different physical properties may enter the parametrisation. The free parameters of the empirical potential are fitted with regard to structure information (e.g. bond length and angle) and energies (e.g. bond strength and reaction energies) as well as thermodynamic properties, e.g. the density of a uni-molecular gas at a certain temperature. The environment of the system may be reflected in the parameters by the choice of an adequate reference set, e.g. a molecule solvated in water. In the following the *optimised potentials for liquid systems*(OPLS) force field is discussed briefly as an example. The OPLS FF was designed originally for organic molecules and peptides, first of all hydro-carbine systems.[12].

Atom Type

The potential energy surface described by the FF can be divided into two parts: the intra and the inter molecular potential. The parameters of the non-bonded or inter molecular potential are classified by the so called atom type for every atom. For the OPLS FF these non-bonded interaction is defined by two center terms depending on the distance $r_{(AB)}$ between two atoms A and B. The first term of the non bonding energy E_{nb} is the Coulomb energy depending on the partial atomic charges q_A and q_B of the atoms. The second term is the Lennard-Jones energy with two parameters per atom, σ_A and ϵ_A, which combines the repulsion between two atoms and the Van der Waals interaction.

$$E_{nb} = \sum_{(AB)} \left[\frac{q_A q_B}{r_{(AB)}} - 4\sqrt{\epsilon_A \epsilon_B} \left(\frac{\sigma_A^3 \sigma_B^3}{r_{(AB)}^6} - \frac{\sigma_A^6 \sigma_B^6}{r_{(AB)}^{12}} \right) \right] \qquad (1.1)$$

Changes in the inter atomic interactions due to polarisation or electronic redistribution are not considered by classical FF, since all atom types are fixed during the simulation. The Lennard-Jones interaction is a quite short ranged interaction term, while the Coulomb interaction only decays with $\frac{1}{r_{(AB)}}$. When using *periodic boundary condition*(pbc) the treatment of infinite interactions is a major issue. One possibility to deal with this situation is the use of a cut-off function, another one is the use of Ewald summation. The latter is superior to a simple cut-off because no artificial boundary is introduced. In 1921, Ewald [13] developed a summation scheme for the potential of infinite (neutral) charge distributions. The sum is then splitted in a long range and a short range interaction part. The short range part is calculated in real space, while the long range interaction is calculated in reciprocal space. The absolute and uniform convergence of the sum is enforced by a factor, e.g. a Gaussian like factor $f(s,x) = e^{-sx^2}$, which can be interpreted as a shielding of the point charges. The factor depends on the parameter s. As the parameter s goes to zero, the convergence factor goes to one, $\lim_{s \to 0} f(s,x) = 1$ and an additional correction term cancels the effect of shielding out in the reziprocal space[14, 15]. A commonly used modification of the Ewald summation is the *particle mesh Ewald approach* (PME)[16] where the calculation in reciprocal space is enhanced by fast Fourier transformation and distances in real space are chosen through an adequately chosen cutoff.

1.1. PARTICLE INTERACTIONS AND FORCES

Topology of the Molecule

The topology of the molecule is defined by a list of all the bonds between two atoms A and B and by the corresponding empirical parameters to describe the bonded or intra molecular interactions, which consists of three terms.
First, the bond term E_{bond} depends on the bond distance $r_{(AB)}$ between each bounded pair of atoms A and B. The bond parameters are therefore defined pairwise, using a harmonic potential defined by a spring constant $k_{(AB)}^{bond}$ and equilibrium distance $r_0^{bond,(AB)}$.

$$E_{bond} = \sum_{(AB)} k_{(AB)}^{bond} \left(r_{(AB)} - r_0^{bond,(AB)} \right)^2 \quad (1.2)$$

Second, the angle term E_{angle} is defined as a quadratic function of the angle $\vartheta_{(ABC)}$ between the bond vector of the bonded atoms A and B and the bond vector of the bonded atoms B and C. The bond angle parameters, i.e. the spring constant $k_{(ABC)}^{angle}$ and the equilibrium angle $\vartheta_0^{angle,(ABC)}$, are therefore defined triple-wise.

$$E_{angle} = \sum_{(ABC)} k_{(ABC)}^{angle} \left(\vartheta_{(ABC)} - \vartheta_0^{angle,(ABC)} \right)^2 \quad (1.3)$$

The third term E_{RB} depends on the dihedral angle, which is defined as the angle between the components of the two bond vectors $\vec{r}_{(AB)}$ and $\vec{r}_{(CD)}$ which is orthogonal to the bond vector $\vec{r}_{(BC)}$, for the atoms A and B, B and C and C and D being bonded to each other, see Eq. 1.4.

$$\begin{aligned}\cos\varphi_{(ABCD)} &= \frac{\left[\vec{r}_{(AB)}\right]^{\perp(BC)} \cdot \left[\vec{r}_{(CD)}\right]^{\perp(BC)}}{r_{(AB)}^{\perp(BC)} r_{(CD)}^{\perp(BC)}} \\ &= \left[\frac{\vec{r}_{(AB)}}{r_{(AB)}} - \frac{\left(\vec{r}_{(AB)} \cdot \vec{r}_{(BC)}\right)\vec{r}_{(BC)}}{r_{(BC)}^2 r_{(AB)}}\right] \cdot \left[\frac{\vec{r}_{(CD)}}{r_{(CD)}} - \frac{\left(\vec{r}_{(CD)} \cdot \vec{r}_{(BC)}\right)\vec{r}_{(BC)}}{r_{(BC)}^2 r_{(CD)}}\right]\end{aligned} \quad (1.4)$$

The potential E_{RB} corresponds to the Ryckaert and Bellman torsional potential, which consists of 5 terms depending on 5 constants $k_n^{(ABCD)}$ that are specified quadruple wise.

$$E_{RB} = \sum_{(ABCD)} \sum_{n=0}^{5} k_n^{(ABCD)} \left(\cos(\varphi_{(ABCD)}) \right)^n \quad (1.5)$$

The non bonded interaction is included in the bonding parameters, therefore it is switched off for atoms being connected to each other via less than 3 bonds.

1.1.2 The Electronic Problem - Ab Initio

The general quantum mechanical description of an atomic system by a time independent Schrödinger equation leads to a coupled differential equation of all particles, atomic nuclei and electrons. As mentioned before, the Born-Oppenheimer approximation allows to describe the motion of electrons in a static external field given by the nuclear positions and charges. In the following, different ab initio quantum mechanical descriptions of the electronic problem are described briefly. The nuclei - as heavy particles - are still considered as classical particles, in the following the coulomb interaction between the nuclei is not mentioned explicitly in the total energy term.

The electronic states are given by the N-electron wave function ψ_e that solves the eigenproblem of the time independent (electronic) Schrödinger equation:

$$\hat{H}_e \psi_e = E_e \psi_e \qquad (1.6)$$

The Hamiltonian operator contains the kinetic energy of the N electrons ($\{i\}$), see term $\hat{H}_{kin}$, the Coulomb interaction with all other electrons ($\{j\}$), see term $\hat{H}_{Coul}^{e-e}$, and the Coulomb interaction with all M nuclei ($\{A\}$) with atomic number Z_A, see term $\hat{H}_{Coul}^{e-n}$ in Eq. 1.7. Below, the Hamiltonian is given in the position space with the distance r_{ij} and the distance r_{iA} between two particles, e.g. electron i and electron j or electron i and nucleus A. The kinetic energy depends on the momentum operator, which is defined by $-i\vec{\nabla}$ in atomic units in the position space, with i being the imaginary unit and $\vec{\nabla}$ describing the 3 dimensional vector of partial derivatives. The spin part of the electronic function is not explicitly given in the formulae, as spin related properties are not the focus of this work.

$$\begin{aligned}\hat{H}_e &= \hat{H}_{kin} + \hat{H}_{Coul}^{e-e} + \hat{H}_{Coul}^{e-n} \\ &= -\sum_{i=1}^{N} \frac{1}{2}\nabla_i^2 + \sum_{i=1}^{N-1}\sum_{j=i+1}^{N} \frac{1}{r_{ij}} - \sum_{i=1}^{N}\sum_{A=1}^{M} \frac{Z_A}{r_{iA}}\end{aligned} \qquad (1.7)$$

Hartree-Fock

In *Hartree-Fock theory*(HF)[17, 18, 19, 20], the many electron system is approximated by a Slater determinant[21] of N orthogonal one-electron wave functions $\{\psi_i\}_N$, which ensures at the same time the antisymmetry with respect to position interchange of two electrons as required by the Pauli principle for fermions.

$$\psi_e = \frac{1}{\sqrt{N!}} \begin{vmatrix} \psi_1(\vec{r}_1) & \psi_1(\vec{r}_2) & ... & \psi_1(\vec{r}_N) \\ \psi_2(\vec{r}_1) & \psi_2(\vec{r}_2) & ... & \psi_2(\vec{r}_N) \\ ... & & & ... \\ \psi_N(\vec{r}_1) & \psi_N(\vec{r}_2) & ... & \psi_N(\vec{r}_N) \end{vmatrix} \qquad (1.8)$$

The kinetic energy ($\hat{H}_{kin}$) and the Coulomb interaction between nuclei and electrons ($\hat{H}_{Coul}^{e-n}$) can be written in terms of one-electron wave functions, as the corresponding operator is a sum of one electron operators:

$$\langle \psi_e | \hat{H}_{kin} | \psi_e \rangle = -\sum_{i=1}^{N} \int \psi_i^\star(\vec{r}_i) \frac{1}{2}\nabla_i^2 \psi_i(\vec{r}_i) d\vec{r}_i = \sum_{i=1}^{N} T_i \qquad (1.9)$$

$$\langle \psi_e | \hat{H}_{Coul}^{e-n} | \psi_e \rangle = -\sum_{i=1}^{N}\sum_{A=1}^{M} \int \psi_i^\star(\vec{r}_i) \frac{Z_A}{r_{iA}} \psi_i(\vec{r}_i) d\vec{r}_i = \sum_{i=1}^{N} U_i$$

The electron-electron interaction $\hat{H}_{Coul}^{e-e}$ involves a mixing of one electronic wave functions. Therefore, the electron-electron interaction is approximated by a mean field potential which depends on the N one electron wave functions $\{\psi_i\}$. Applying the Slater Condon rule, the following two terms are obtained:

$$\langle \psi_e | \hat{H}_{Coul}^{e-e} | \psi_e \rangle = \frac{1}{2}\sum_{i=1}^{N}\sum_{j=1}^{N}[J_{ij} - K_{ij}] \qquad (1.10)$$

1.1. PARTICLE INTERACTIONS AND FORCES

In Eq. 1.10, the Coulomb term corresponds to the two electron interaction matrix J_{ij} in the $N \times N$ dimensional one-electron vector space:

$$J_{ij} = \int d\vec{r}_j d\vec{r}_i \psi_j^\star(\vec{r}_j)\psi_i^\star(\vec{r}_i)\frac{1}{r_{ij}}\psi_j(\vec{r}_j)\psi_i(\vec{r}_i) \quad (1.11)$$

The second term K_{ij} in Eq. 1.10 is the exchange term which compensates the Coulomb self interaction ($i = j$) and maintains the Pauli principle at the exchange of two electrons.

$$K_{ij} = \int d\vec{r}_j d\vec{r}_i \psi_j^\star(\vec{r}_i)\psi_i^\star(\vec{r}_j)\frac{1}{r_{ij}}\psi_j(\vec{r}_j)\psi_i(\vec{r}_i) \quad (1.12)$$

The total energy evaluates as follows in the $N \times N$ - matrix (E_{ij}) formulation:

$$E[\psi] = \sum_{i=1}^{N}\sum_{j=1}^{N}\left[T_i\delta_{ij} + U_i\delta_{ij} + \frac{1}{2}J_{ij} - \frac{1}{2}K_{ij}\right] = \sum_{j=1}^{N}\sum_{i=1}^{N}E_{ij} \quad (1.13)$$

The one-electron functions are defined by a linear expansion of basis functions ϕ_ν with expansion coefficients $c_{i\nu}$. Atomic orbitals are often used as basis functions. This is the so called *linear combination of atomic orbital approach*(LCAO). The atom centred local basis functions $\phi_\nu = \phi_i^A$ are defined as a minimal basis of orthogonal functions relating to the same atom, but the complete set of atomic orbitals $\{\phi_i^A, \phi_j^B, ...\}$ is not a minimal basis nor orthogonal.

$$\psi_i = \sum_\nu c_{i\nu}\phi_\nu \quad (1.14)$$

$$E_{ij} = \sum_\mu \sum_\nu c_{\mu i}^\star c_{j\nu} E_{\mu\nu} \quad (1.15)$$

The matrix $E_{\mu\nu}$ corresponds to the matrix elements of Eq. 1.13 evaluated with the LCAO basis. Employing the variational principle to Eq. 1.13 and diagonalising the electronic problem by a unitary transformation, one obtains the following matrix equations (Eq. 1.18) for the unknown expansion coefficients $c_{i\nu}$[19, 20]:

$$\frac{\delta}{\delta\phi_\mu}\left[E[\psi] - \sum_{i=1}^{N}\epsilon_i \int d\vec{r}\psi_i^\star\psi_i\right] = 0 \quad (1.16)$$

$$\frac{\delta}{\delta\phi_\mu}\left[\sum_{j=1}^{N}\sum_{i=1}^{N}\sum_{\nu=1}^{}\sum_{\mu=1}^{}c_{\mu j}^\star c_{i\nu}E_{\nu\mu} - \sum_{i=1}^{N}\sum_{\nu=1}^{}\sum_{\mu=1}^{}\epsilon_i c_{\mu j}^\star c_{i\nu}\int d\vec{r}\phi_\mu^\star\phi_\nu\right] = 0 \quad (1.17)$$

$$\sum_{\nu=1}^{K}c_{i\nu}[E_{\mu\nu}(\{c_{\eta k}\}) - \epsilon_i S_{\mu\nu}] = 0 \quad (1.18)$$

With the overlap matrix $S_{\mu\nu} = \int d\vec{r}\phi_\mu^\star\phi_\nu$. The secular equation 1.18 is solved iteratively as $E_{\mu\nu}$ depends on the coefficients; the converged calculation is referred to as a *self consistent field calculation* (SCF).

The so called post-HF methods are an advancement of HF method in order to obtain a better description of the electronic correlation, e.g. the Møller-Plesset perturbation theory[22]. The Møller-Plesset perturbation theory to the second order term is called MP2 method.

Density Functional Theory

A different approach was developed by Hohenberg and Kohn [23, 24, 25]. The so called *density functional theory (DFT)* was founded on the statement that the energy $E[\rho]$ - as any other property of the system - is a unique functional of the electron density of the ground state ρ_0. The computational advantage is to replace the (3N)-dimensional eigen-problem of the N electrons system by a three-dimensional problem depending on the overall density of electrons. The variational principle can be applied in this case, since the ground state energy is lower than the energy belonging to any other density distribution of the many particle system ($E[\rho] \geq E[\rho_0]$), see Eq. 1.19.

$$0 = \frac{\delta}{\delta \varrho}\left(E[\varrho] - \mu \left(\int \varrho(\vec{r})d\vec{r} - N\right)\right) \quad (1.19)$$

In general, the electron density is defined as the (N-1) dimensional integral over the electronic wave function ψ_e. Introducing the Kohn-Sham theorem, the ground state of a system of interacting electrons is equivalent to the ground state of a special non interacting system. This statement corresponds to the mean field approximation in the Hartree-Fock theory. Thereby, the electron density is expressed in terms of N orthogonal one electron functions (Eq. 1.8), see Eq. 1.20.

$$\varrho(\vec{r}) = \sum_{i=1}^{N} \int \psi_e^\star(\vec{r}_1,...,\vec{r}_N)\psi_e(\vec{r}_1,...,\vec{r}_N)\delta(\vec{r} - \vec{r}_i)d\vec{r}_1...d\vec{r}_N = \sum_{i=1}^{N} |\psi_i|^2 \quad (1.20)$$

To obtain a useful expression for the energy functional, at first the energy U_i (see Eq. 1.10) is expressed as a functional of the electron density as defined in Eq. 1.20, which results in the functional of the external potential $E_{ext}[\varrho]$

$$\sum_i U_i = -\sum_{A=1}^{M} \int \left[\sum_{i=1}^{N} \psi_i^\star(\vec{r}_i)\psi_i(\vec{r}_i)\right] \frac{Z_A}{r_{iA}} d\vec{r}_i \quad (1.21)$$

$$E_{ext}[\varrho] = -\sum_{A=1}^{M} \int [\varrho(\vec{r}_i)] \frac{Z_A}{r_{iA}} d\vec{r}_i \quad (1.22)$$

The same transformation is performed for the term J_{ij} (see Eq. 1.11), which is named electronic Coulomb functional $E_{Coul}[\varrho]$. The squared bracket mark here the electronic density expression.

$$\sum_{ij} J_{ij} = \frac{1}{2}\int d\vec{r}_j d\vec{r}_i \left[\sum_j \psi_j(\vec{r}_j)\psi_j^\star(\vec{r}_j)\right] \frac{1}{r_{ij}} \left[\sum_i \psi_i^\star(\vec{r}_i)\psi_i(\vec{r}_i)\right] \quad (1.23)$$

$$E_{Coul}[\varrho] = \frac{1}{2}\int d\vec{r}_j d\vec{r}_i [\varrho(\vec{r}_j)] \frac{1}{r_{ij}} [\varrho(\vec{r}_i)] \quad (1.24)$$

The kinetic energy E_{kin}, Eq. 1.9, and the exchange-correlation term, Eq. 1.12, are not transformed directly to density functionals, but the total energy can formally be written as follows.

$$E[\varrho] = E_{kin} + E_{ext}[\varrho] + E_{Coul}[\varrho] + E_{xc}[\varrho] \quad (1.25)$$

As in the HF, the kinetic energy in DFT is calculated by an effective one-electron Schrödinger operator, Eq. 1.26, with the density as defined in Eq. 1.20 and the kinetic energy as defined in Eq. 1.9 depending on the one-electron wave functions ψ_i.

$$\epsilon_i \psi_i = (\hat{T} + \hat{V}_{eff}(\varrho))\psi_i \quad (1.26)$$

1.1. PARTICLE INTERACTIONS AND FORCES

In this formula, the kinetic operator $\hat{T}$ and the effective potential operator $\hat{V}_{eff}$ were introduced.

$$\hat{T} = -\frac{1}{2}\sum_i \nabla_i^2 \quad (1.27)$$

$$\hat{V}_{eff} = \hat{V}_{Coul}(\varrho) + \hat{V}_{ext} + \hat{V}_{XC} \quad (1.28)$$

The latter consists of all the non kinetic parts of the energy operator, i.e. the electronic Coulomb interaction (Eq. 1.29), the energy from the external field, which means the Coulomb interaction with the atomic nuclei, (Eq. 1.30) and the (still unknown) exchange-correlation term $\hat{V}_{XC}$.

$$\hat{V}_{coul}(\varrho) = \int \frac{\varrho(\vec{r}')}{|\vec{r}-\vec{r}'|}d\vec{r}' \quad (1.29)$$

$$\hat{V}_{ext}(\varrho) = \sum_A \frac{Z_A}{|\vec{r}-\vec{r}_A|} \quad (1.30)$$

The exchange-correlation term $\hat{V}_{XC}$ (or $E_{xc}[\varrho]$) needs further approximation. The ideal exchange-correlation functional contains the differences to the exact kinetic and electron-electron interaction energies and it should in this way compensate (all) errors of the approximation. Many different functionals have been developed in the past, such as the *local density* (LDA) and the *generalised gradient* (GGA) approximation, as well as hybrid methods. The latter are linear combinations of the HF exchange energy and of the known DFT exchange correlation functionals, such as LDA and GGA. The weighting factor enters as an empirical parameter. The pioneer of the hybrid approach was Becke[26]. His efforts resulted in the famous Becke-three-parameter-Lee-Yang-Parr functional (B3LYP).

The computational scheme results as follows. The wave functions ψ_i are obtained as solutions of a one particle Hamilton operator (see Eq. 1.26). The Hamilton operator depends on the electron density ϱ that itself depends on the wave functions ψ_i, see Eq. 1.20. Therefore, the DFT equations are solved iteratively until convergence to maintain the self consistent field (SCF). For DFT calculations analogue to the Hartree Fock method, the equations are solved using a linear combination of basis functions, usually atomic basis sets or plane wave functions, as an ansatz of the calculation in order to obtain matrix equations for the coefficients.

Apart from the exchange-correlation functional, the choice of the basis function influences the DFT calculation. The use of an extended basis set increases the computational time of the calculation, but a too small basis set will lead to a wrong solution. Many basis sets provide additional so called polarisation functions to the minimal basis of each atom, which are asymmetric auxiliary functions. The so called Popel basis sets denote the polarizability by the symbol *, e.g. 6-31g*. The influence of the number of basis function is usually analysed by a systematic increase of the basis set to induce convergence to the complete basis set limit. The notation double, triple or even higher tuple zeta denotes the use of 2, three or more functions per (valence) atomic orbital. One example are the correlation-consistent polarised valence basis sets, denoted cc-pVDZ, cc-pVTZ, cc-pVQZ.

1.1.3 The Electronic Problem - Density Functional based Tight Binding

The Density Functional based Tight Binding method (DFTB)[27, 28] is an approach that was developed from DFT in the mid 1990s. The aim of the method is to reduce computational costs for extended system and time scales by introducing tight binding approximations and taking precalculated parameters from two center interaction lists.

Starting from the DFT formulation of the electronic density, Eq. 1.20, ϱ is approximated by the starting density ϱ_0 and density fluctuations as introduced by Foulkes and Haydock [29]. This leads to an approximation of the energy functional, Eq. 1.25, neglecting higher than quadratic terms in density fluctuations.

$$\varrho = \varrho_0 + \delta\varrho \tag{1.31}$$
$$E(\varrho) = \underbrace{E_{BS}(\varrho_0) + E_{rep}(\varrho_0)}_{\text{zeroth order}} + \underbrace{0}_{\text{linear order}} + \underbrace{E_{2nd}(\varrho_0, \delta\varrho)}_{\text{second order}} + \underbrace{\ldots}_{\text{higher order}} \tag{1.32}$$

The linear order of the expansion is zero due to symmetry considerations. A limitation of the method to the **zeroth order** corresponds to a loss of self consistency - in the sense of SCF in DFT - as no iterative solving is performed in this case. The zeroth order is divided into two terms.

$$E_{BS}(\varrho_0) = \sum_{i=1}^{N} \langle \psi_i \mid \hat{T}(\varrho_0) + \hat{V}_{eff}(\varrho_0) \mid \psi_i \rangle \tag{1.33}$$

$$E_{rep}(\varrho_0) = -\frac{1}{2} \int \int \frac{\varrho_0(\vec{r})\varrho_0(\vec{r'})}{|\vec{r'}-\vec{r}|} d^3\vec{r'}d^3\vec{r} + E_{xc}(\varrho_0) - \int V_{xc}(\varrho_0)\varrho_0 d^3\vec{r} \tag{1.34}$$

The first is the so called band structure term, the second is the repulsive term. The band structure term depends on electronic wave functions as in DFT, see Eq. 1.26. The repulsive term is an additional short ranged correction to this term. As the method uses a tight binding approach, the starting density is approximated by a superposition of the electron density of neutral atoms $\{A\}$ - a pseudo-atomic density, see Eq. 1.35, and the wave functions are defined as a LCAO, see Eq. 1.14.

$$\varrho_0(\vec{r}) = \sum_A \varrho_0^A(\vec{r} - \vec{R}_A) \tag{1.35}$$

The wave functions are noted as ϕ_μ^A, where the capital letter signifies the atomic nucleus - reference point of the function - and the Greek letter means the atomic orbital. Compact Slater type functions $\phi_\mu^A(\vec{r})$ are used as orbitals with atom dependent constants.[30, 31]

$$\phi_\mu^A(\vec{r}) = \phi_{nlm}^A = F_{ml}(r) \, Y_{lm}\left(\frac{\vec{r}}{r}\right) \tag{1.36}$$

$$F_{ml}(r) = \sum_{i=1}^{5}\sum_{j=0}^{3} a_{ij} r^{l+j} e^{\alpha_i r} \tag{1.37}$$

where Y_{lm} is the real spherical harmonic and F_{ml} signifies the spherically symmetric part of the function. The constants are determined by solving the one electron mean field atomic Schrödinger equation (Eq. 1.26) modified by the additional term $(r^2 \, r_0^{-2})$ in order to obtain more compressed functions[32]. The factor r_0 is chosen to be about 1.85 times the covalent radius[i]. The five parameters of α_i in Eq. 1.37 depending on the atom type and j going from zero to three is a sufficiently accurate basis set for all elements up to the third row[32].

---

[i]In earlier publications also twice the covalent radius is found for r_0[28] and depending on the parameterisation, the factor r_0 deviates from the estimate of 1.85.

1.1. PARTICLE INTERACTIONS AND FORCES

Based on the tight binding approach, both zeroth order terms of Equation 1.34 are approximated in the following by two particle interaction terms, which leads to terms that are characterised only by the atomic orbitals $\phi_\mu^A(\vec{r})$, the starting density ϱ_0 and the distance between the atomic nuclei. These parameters are preestimated and all integrals are precalculated and listed in biatomic interaction lists, which are called SK-files referring to the integral tables for the band structure which was introduced by Slater and Koster.

Firstly, using a two centre approximation ($\langle \phi_\nu^A \mid \hat{V}_{eff}(\varrho_0) \mid \phi_\mu^B \rangle \approx \langle \phi_\nu^A \mid \hat{V}_{eff}(\varrho_0^A) + V_{eff}(\varrho_0^B) \mid \phi_\mu^B \rangle$), the band structure term H^{AB} transforms in the following matrix elements, that only depend on two interacting atoms $\{AB\}$ and will be calculated in advance and tabulated in the SK-files per orbital and distance of the nuclei A and B.

$$E_{BS}(\varrho_0) = \sum_{i=1}^{N} \sum_{AB}^{M} \sum_{\nu\mu} c_{\mu i}^A c_{\nu i}^B [H^0]_{\nu\mu}^{AB} \tag{1.38}$$

$$[H^0]_{\nu\mu}^{AB} = \langle \phi_\nu^A \mid \hat{T}(\varrho_0) + \hat{V}_{eff}(\varrho_0) \mid \phi_\mu^B \rangle \tag{1.39}$$

$$\approx \begin{cases} \langle \phi_\nu^A \mid \hat{T} + \hat{V}_{eff}^A(\varrho_0^A) + \hat{V}_{eff}^B(\varrho_0^B) \mid \phi_\mu^B \rangle & A \neq B \\ 0 & A = B; \nu \neq \mu \\ \langle \phi_\nu^A \mid \hat{T} + \hat{V}_{eff}^A(\varrho_0^A) \mid \phi_\nu^A \rangle = E_\nu^A & A = B; \nu = \mu \end{cases}$$

Secondly, the repulsive potential is approximated in a similar way by a cluster expansion, where only pair interactions are considered, see Eq. 1.40.

$$E_{rep}(\varrho_0) = \sum_A E_{rep}(\varrho_0^A) + \frac{1}{2} \sum_{AB} U_{rep}(\varrho_0^A, \varrho_0^B) + ... \tag{1.40}$$

In DFTB, as a semi empirical method, repulsive potential (E_{rep}) is parametrised by DFT calculations for test system set $\{AB\}$. One has to mention that - as for all semi empirical methods - the fitting procedure, the choice of the test set and of course also the reference (DFT) method influence the parameters significantly and need to be chosen with care. For each scientific problem, a preexisting SK-file parameter set should be at least validated. The obtained repulsive potential $E_{rep}^{(SK)}$ is also tabulated in the SK-files only depending on the distance R_{AB} of the nuclei A and B.

$$E_{rep}^{(SK)}(R_{AB}) \approx \left[E_{tot}^{(DFT)} - E_{BS}(\varrho_0) - E_{2nd}(\varrho_0, \delta\varrho) \right] \quad \forall \{AB\} \tag{1.41}$$

Second Order Term

The second order term[33, 28], see Eq. 1.42, consists of a long range Coulomb like term and a short range exchange-correlation term.

$$E_{2nd}(\varrho_0, \delta\varrho) = \frac{1}{2} \int \int \left[\frac{1}{|\vec{r}' - \vec{r}|} + \frac{\delta^2 E_{xc}}{\delta\varrho\delta\varrho'} \right] \delta\varrho\delta\varrho' \tag{1.42}$$

The density fluctuations per atom $\delta\varrho_A$ are expanded as series of radial and angular functions ($\delta\varrho_X = \sum k_{ml} F_{ml}^X Y_{ml}$), see Eq. 1.37, but only the monopole like distribution with a exponential decay is taken into account ($\delta\varrho_x \approx \Delta q F_{00}^X Y_{00}$), which leads to a term depending on changes (Δq_A) in the Mulliken population (q_A), which is defined in Eq. 1.44, and the factor

Γ_{AB} (Eq. 1.43).

$$E_{2nd} \approx \frac{1}{2} \sum_{AB} \Gamma_{AB}(U^A, U^B, R_{AB}) \Delta q_A \Delta q_B \qquad (1.43)$$

$$q_A = \frac{1}{2} \sum_i^N \sum_B \sum_{\nu\mu} \left(c^A_{\mu i} c^B_{\nu i} \langle \phi^A_\mu | \phi^B_\nu \rangle + c^B_{\nu i} c^A_{\mu i} \langle \phi^B_\nu | \phi^A_\mu \rangle \right) \qquad (1.44)$$

Γ_{AB} in Eq. 1.43 is approximated as a function depending on the distance between two nuclei and the Hubbard parameter U^A (or chemical hardness) and is usually written in the following functional form[28]:

$$\Gamma_{AB} = \frac{1}{R_{AB}} - X_{AB}(U^A, U^B, R_{AB}) \qquad (1.45)$$

Here, the factor X_{AB} in Eq. 1.45 is set with respect to the limits of Eq. 1.43. In the long range regime, ($|R_A - R_B| \to \infty$), the Coulomb interaction is the dominant term in Γ_{AB} as mentioned above. In the short range case, $|R_A - R_B| \to 0$, Γ_{AB} equals the Hubbard parameter or chemical hardness ($U_A = U_B$), which is an approximation to the self interaction contribution[i].

Hydrogen Bond Interaction

In order to enhance the description of hydrogen bond interaction, the function Γ_{AB} is modified for interaction of hydrogen atom A with heavy atom B or *vice versa*, the short range term being damped by an exponential factor that introduces an additional set of (empirical) parameters[34]. This DFTB modification is called H-bond-DFTB in the following.

$$\Gamma^{(H-bond)}_{AB} = \frac{1}{|R_{AB}|} - X_{AB} \exp\left[-\left(\frac{U_A + U_B}{2}\right)^4 |R_{AB}|^2\right] \qquad (1.46)$$

By damping the short range part, Γ_{AB} becomes more repulsive especially in the region of covalent bond (0.1-0.2 nm), which leads to larger polarisation of the respective polar bond and thereby improves the hydrogen bonding in particular[35]. To solve the DFTB equations, Eq. 1.32, the variational principle is applied and the functional derivative of the total energy (Eq. 1.49.) is calculated with respect to the density under the condition of normalised basis function. This results in the secular equation, valid for all A, B and ν, which is the matrix equation for the constant $c^A_{\mu i}$.

$$0 = \sum_\mu c^A_{i\mu} \left(H^{AB}_{\mu\nu} - \epsilon_i S^{AB}_{\mu\nu} \right) \quad \forall A, B, \nu \qquad (1.47)$$

$$H^{AB}_{\mu\nu} = [H^0]^{AB}_{\mu\nu} + [H^{2nd}]^{AB}_{\mu\nu} \qquad (1.48)$$

$$E(\varrho) = E_{BS}(\varrho_0) + E_{rep}(\varrho_0) + E_{2nd}(\varrho_0, \delta\varrho) \qquad (1.49)$$

The repulsive potential does not enter the secular equation, as it is a constant regarding the energy density fluctuations or $c^A_{\mu i}$. Considering the second order term, the equations are solved iteratively, i.e. self consistent charge DFTB. Otherwise the correction of the band structure Hamiltonian is neglected in Eq. 1.48, $H^{AB}_{\mu\nu} \approx [H^0]^{AB}_{\mu\nu}$, as well as the total energy lacks the second order term E_{2nd} in Eq. 1.49.

---

[i]This approach is commonly used for semi empirical methods

1.1. PARTICLE INTERACTIONS AND FORCES

Intermolecular Forces

So far, DFTB describes the total energy and the electronic configuration of an atomic system with fixed nuclear positions. For a molecular dynamics simulation, as discussed in Sec. 1.2, the internuclear forces need to be calculated. They are derived from Eq. 1.49 as the derivative of the total energy with respect to the nucleus coordinates, for details see Ref. [28].

$$\vec{F}^A = -\sum_{i=1}^{N} \sum_{AB} \sum_{\mu\nu} c_{\mu i}^A c_{\nu i}^B \left(\frac{\partial H_{\mu\nu}^{AB}}{\partial \vec{R}_A} - \left[\epsilon_i - \frac{[H^{2nd}]_{\mu\nu}^{AB}}{S_{\mu\nu}^{AB}}\right] \frac{\partial S_{\mu\nu}^{AB}}{\partial \vec{R}_A} \right)$$
$$-\Delta q_A \sum_B \frac{\partial \Gamma_{AB}}{\partial \vec{R}_A} \Delta q_B - \frac{\partial E_{rep}}{\partial \vec{R}_A}$$

1.1.4 The QM/MM Approximation

In case of localised changes in the electronic structure, due to chemical reactions or charge transfers, a common approach is the coupling of quantum methods to empirical force field methods[36] [37] [38]. The system is divided into two regions; the QM region, where chemical reactions take place, and the MM region, that contains the chemical environment for the reaction in question, but does not experience any electronic-structural changes. In the last few decades, several approaches of QM/MM coupling have been developed especially for biological systems. An overview is given by Senn et al. [39]. In the following, only the used approach is presented.

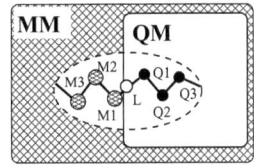

Figure 1.1: *QM/MM coupling scheme; QM-zone: Q3, Q2, Q1 and Link Atom (L); MM-zone: M3, M2, M1.*

In the so called additive scheme, the Hamiltonian is divided into three parts; the first part describes the interaction between atoms inside the MM zone, the second part contains the interactions between atoms inside the QM zone and the third term contains all other interactions, i.e. the interactions which involve atoms from the QM and the MM zone.

$$H_{total} = H_{MM} + H_{QM} + H_{QM/MM} \tag{1.50}$$

As electronic embedding is used, the MM charges are included in the QM calculation as point charge background, which leads to a polarisation of the QM zone. More sophisticated schemes include polarisation of the force field atoms, which is of course highly dependent on the choice of the FF parameters needed to express the polarisation effect. In case a covalent bond is cut by the QM/MM boundary, the dangling bond is coupled by a so called *link atom* (L). Therefore, for each covalent bond between a QM atom (Q1) and a MM atom (M1) (see Fig. 1.1) a link atom (L) is introduced. It is treated like a hydrogen atom in the QM zone. There is no interaction between the link atom and atoms in the MM zone. Via constraints, the artificial degrees of freedom are removed from the extended system and the forces acting on (L) are distributed onto (M1) and (Q1), which are kept at a certain distance and angle with respect to the link atom (L). To avoid an over polarisation of the QM atoms at the boundary, due to the proximity of Q1 and L to M1, the charge of M1 may be distributed among the other atoms of the same charge group.

In general the MM and QM zone should be neutral, or at least contain multiples of the electronic charge e. The cut covalent bond should be symmetric and not polar, e.g. a carbon-carbon bond. In the literature, one also finds methods that allow to place a QM/MM boundary in a polar bulk material[40].

For the present study, the QM/MM implementation of Gromacs program[41] was modified in order to introduce a coupling between Gromacs and DFTB+ code[42]. The details are given in the appendix B.1. The actual additive schema is divided between the two programs as follows:

$$H_{total} = \left[H_{MM}^{total} + H_{QM/MM}^{bonded} + H_{QM/MM}^{LJ}\right]_{Gromacs} + \left[H_{QM+L}^{total} + H_{(QM+L)/MM}^{ele}\right]_{DFTB+} \quad (1.51)$$

At each time step, the coordinates and atom types of the QM zone (plus L) and the distribution of point charges (MM - zone) are handed to DFTB+, where a scc-DFTB calculation is performed. The total energy of the subsystem (latter term in Eq. 1.51) and the according forces are calculated, forces on the QM atoms as well as the forces on the MM atoms due to electrostatic interaction with the QM atoms.

All other forces are taken from the Force Field description, see left picture of Fig. 3.7. The MD step is performed as described in Sec. 1.2 regarding all forces, those obtained from the QM calculation and those from the FF calculation.

1.2 Classical Molecular Dynamics Simulations

After comments on particle interactions and forces, the thermodynamic aspects of a many particle system in simulations are described in the following. The classical description of *atomistic molecular dynamics* (MD) obeys the Newtonian equation of motion applied to the centre of mass (m_A) of all atoms ($\{A\}$), which are explicitly introduced in the simulation by their three dimensional ($a \in [1, 2, 3]$) coordinates (x_a^A), velocities (v_a^A) and forces (F_a^A). Since only atomic forces are considered, the total energy is conserved and the equations of motion are solved for a given interaction by any numerical scheme, e.g. leap frog verlet integration[43]. Considering a fixed system size (see below, section 1.1.1), a many particle system described by MD simulation corresponds - thermodynamically speaking - to the micro canonical ensemble (NVE). In the following, important thermodynamic properties and *periodic boundary conditions* (pbc) are discussed. The adequate description of the equations of motion by a Hamiltonian formalism is introduced. Based on this, changes in the equations of motion by thermostats and barostats are pointed out, which enables the description of different thermodynamic ensembles, such as the *canonical ensemble* (NVT) and an *ensemble with constant number of particles, pressure and temperature* (NPT). Thermodynamic properties in the simulation are defined in the context of statistical thermodynamics.

1.2.1 Thermodynamic properties

The instantaneous microscopic properties follow directly from the simulation, e.g. positions (x_a^A), velocities (v_a^A) and forces (F_a^A) of all particles or atoms as well as the potential (E_{pot}) and the kinetic energy (E_{kin}). Structural properties such as the radial distribution function(RDF) only depend on the coordinates.

The RDF is a one dimensional function which is defined between two groups of atoms or species with the total number of elements being N_I and N_J. Each element of group I or J

1.2. CLASSICAL MOLECULAR DYNAMICS SIMULATIONS

is chacterised by its coordinates $\vec{R}_i$ or $\vec{R}_j$.

$$\text{RDF} = g(r) = \frac{V}{N_I N_J} \sum_{j}^{N_J} \sum_{i}^{N_I} \delta(r - |\vec{R}_i - \vec{R}_j|) \tag{1.52}$$

$$N = 4\pi \rho_J \int_0^R r^2 g(r) dr \tag{1.53}$$

The integral over the RDF (Eq. 1.53) defines the average number of elements of group J in a sphere with radius R around each element of group I, the so called number density. The density of elements of group J in the system is denoted ρ_J.

In contrast to the microscopic properties, thermodynamic properties are defined as averaged properties by means of statistical thermodynamics. In a simulation, the instantaneous temperature is defined as the kinetic temperature $(T = \frac{2}{k_B N_f} E_{kin})$[i], as known from the equipartition of energy[44].

Tensorial Virial Theorem

The tensorial virial theorem provides a useful expression for pressure (P) during the simulation. To derive the formulae as proposed by Park [45], the equations of motion are multiplied by the space coordinate x_b^A and the oncoming terms are physically interpreted.

$$\Sigma_{ba} = 3VP_{ba} - 2E_{ba} \tag{1.54}$$

Here, three tensors are introduced, which are defined as follows:

$$\Sigma_{ba} = \sum_A x_b^A F_a^A \tag{1.55}$$

$$P_{ba} = \sum_A \frac{1}{3V} \frac{d}{dt}(m x_b^A \frac{dx_a^A}{dt}) \tag{1.56}$$

$$E_{ba} = \sum_A \frac{m^A}{2} \frac{dx_b^A}{dt} \frac{dx_a^A}{dt} \tag{1.57}$$

The virial tensor Σ_{ba} is related to the interaction potential Φ in the case of forces derived from a potential.

$$\Sigma_{aa} = -\sum_A x_a^A \frac{d\Phi}{dx_a^A} \tag{1.58}$$

The symmetric term E_{ab} and the term P_{ab} correspond to tensorial expressions of the kinetic energy (E_{kin}) and the pressure (P) which are evaluated as the traces of the matrices:

$$E_{kin} = \sum_a E_{aa} = \frac{1}{2} \sum_a \sum_A \left(\frac{dx_a^A}{dt}\right)^2 m^A \tag{1.59}$$

$$P = \sum_a P_{aa} = \frac{1}{3V} \Sigma_{aa} + \frac{2}{3V} E_{aa} \tag{1.60}$$

To understand the expression for pressure P, particles in a quadratic volume ($V = L^3$) are considered. Explicitly introducing the box length L in the coordinates, the inner energy which

---
[i]k_B - Boltzmann constant, N_f - number of degree of freedom

corresponds to the micro canonical Hamiltonian H_{mc} is formulated in terms of $(r_i = \frac{x_a^A}{L})$ and $(y_i = Lp_a^A = m^A L \frac{dx_a^A}{dt})^i$.

$$H_{mc} = E_{kin}(\{\vec{p}^A\}) + \Phi(\{\vec{x}^A\}) = \sum_A \frac{(\vec{p}^A)^2}{2(m^A)} + \Phi(\{\vec{x}^A\}) \quad (1.61)$$

$$= \sum_i \frac{y_i^2}{2L^2 m^i} + \Phi(\{Lr_i\})$$

The micro canonical Hamiltonian $H_{mc}(\{\vec{p}^A\}, \{\vec{x}^A\})$ is expressed in terms of the generalised coordinates x_a^A and $p_a^{A\text{ii}}$, see Formula 1.62. $H_{mc}(\{\vec{p}^A\}, \{\vec{x}^A\})$ provides an equivalent description of the Newtonian equations of motion assuming the existence of an adequate potential Φ.

$$-\frac{dH}{dx_a^A} = \frac{dp_a^A}{dt} \quad \rightarrow \quad -\nabla\Phi(\{\vec{x}^A\}) = \frac{d\vec{p}^A}{dt} \quad (1.62)$$

$$\frac{dH}{dp_a^A} = \frac{dx_a^A}{dt} \quad \rightarrow \quad p_a^A = m^A \frac{d\vec{x}^A}{dt}$$

Pressure in statistical thermodynamics is defined as the negative of the derivative of the thermodynamic potential - inner energy (E) - with respect to the volume, $P = -\left(\frac{dE}{dV}\right)|_{NS}$. The pressure obtained by this definition equals the definition that follows from the virial equation (compare Formula 1.60 and 1.63).

$$P = -\frac{dH(r_a, y_a)}{dV} = -\frac{dH(r_a, y_a)}{dl}\frac{V^{-2/3}}{3} \quad (1.63)$$

$$= -\frac{V^{-2/3}}{3}\left(-2l^{-3}\sum_i \frac{y_i^2}{2m^i} + \sum_i \frac{r_i}{l}\frac{d\Phi(\{lr_i\})}{dr_i}\right)$$

$$= \frac{1}{3V}\left(2\sum_i \frac{m^A}{2}\left(\frac{dx_a^A}{dt}\right)^2 - \sum_{A,a} x_a^A \frac{d\Phi}{dx_a^A}\right)$$

Computer simulations are restricted to finite dimensions, thus to a finite volume V, which is given by the so called simulation box (or supercell) described by the box vectors (b_a). To avoid surface effects, adequate energy conserving *periodic boundary conditions* (pbc) have to be introduced. The commonly used pbc attach to every particle its infinite number $(l, m, n \in \mathbb{Z})$ of periodic images, which means that the original coordinates (x_a^A) are shifted by an integer multiple of the box vectors, $(x_a^{A'} = x_a^A + lg_a^1 + mg_a^2 + ng_a^3)$. Even though the system is thereby surfaceless, the box size is still a sensible parameter in the simulation. Boundary effects such as artificial self interaction of the particles with their periodic images are avoided by using sufficiently large systems.

1.2.2 Different Thermodynamic Ensembles

As already mentioned above, *classical Molecular Dynamics* (MD) simulations as described so far correspond to the *micro canonical ensemble* (NVE). The thermodynamic potential of this ensemble is the entropy S, which is minimal under equilibrium conditions. In many processes, the system is not thermodynamically isolated, it rather experiences changes in energy or

---
[i] i going from 1 to 3N, a going from 1 to 3, A going from 1 to N
[ii] For a constant factor L also y_i and r_i are generalised coordinates.

1.2. CLASSICAL MOLECULAR DYNAMICS SIMULATIONS

volume. Often used conditions lead to a *canonical ensemble* (NVT) or an *isothermal-isobaric ensemble* (NPT). Different thermostatic or barostatic methods are available to maintain a temperature or pressure conserving system, respectively. Among these methods are the Nosé-Hoover Thermostat and the Parrinello-Rahman Barostat and the Berendsen Thermostat and Barostat, that are mainly applied in this study.[14]

Nosé - Hoover Thermostat

The widely used Nosé - Hoover Thermostat[46] achieves a correct description of the canonical distribution. In contrast to the Anderson thermostat[47], where a Maxwell distribution of the velocities is achieved by introducing collisions with a fictitious (not explicitly simulated) heat bath, the Nosé-Hoover method defines the heat bath through a pair of generalised (Lagrangian) coordinates, a space coordinate ($\vec{x}^A$) and its corresponding momentum ($\vec{p}^A = m^A \frac{d x^A}{dt}$), see Eq. 1.62. The isothermal conditions are obtained by scaling the time coordinate (t) of the simulation with the parameter s, which belongs to a new set of generalised coordinates. Therefore, additional terms are introduced in the Hamiltonian and the kinetic energy term is divided by the scaling factor, compared with the microcanonical Hamiltonian, Eq. 1.61.

$$H = H_{Nose} = \sum_j \frac{(p'_j)^2}{2(m_j)s^2} + \Phi(x'_j) + \psi k T_0 \ln(s) + \frac{(f)^2}{2Q} \tag{1.64}$$

The transformation of time is given by ($s dt = dt'$, $p^A_a = p'_i$, $x^A_a = x'_i$)^i and the new (generalised/canonical) corresponding coordinates are (x'_i, p'_i) and (s, f). According to the Hamilton Lagrange formalism, relations between the coordinates $v = \{x'_i, s\}$ and corresponding momenta $w = \{p'_i, f\}$ are defined; $\dot{v} = \frac{dH}{dw}$ and $\dot{w} = -\frac{dH}{dv}$. Nosé proved the canonical distribution of the coordinates x^A_a and p^A_a/s.

$$s \frac{d}{dt'} x'_j = \dot{x}_j = s \frac{dH}{dp'_j} = \frac{p'_j}{(m_j)s} \qquad s \frac{d}{dt'} p'_j = \dot{p}_j = -s \frac{dH}{dx'_j} = -s \frac{d\Phi(x'_j)}{dx'_j} \tag{1.65}$$

$$s \frac{d}{dt'} s = \dot{s} = s \frac{dH}{df} = \frac{sf}{Q} \qquad s \frac{d}{dt'} f = \dot{f} = -s \frac{dH}{ds} = \sum_j \frac{(p'_j)^2}{(m_j)s^2} - \psi k T$$

To avoid time scaling, Hoover introduced the friction coefficient $\zeta^{fric} = f/Q$ (and $d\zeta^{fric} = df/Q$) and reformulated the equations of motion. This set of variables does not represent a new set of generalised coordinates as no adequate Hamiltonian is found, but the statistical distribution of x^A_a is not affected by the redefinition and for the practical use it is advantageous to avoid time scaling. The equation of motion below follows from the equations 1.65 after reformulation in the new coordinates:

$$\ddot{x}^A_a = \frac{d}{dt} \dot{x}^A_a = -\frac{d\Phi(x^A_a)}{dx^A_a} \frac{1}{(m^A)} - \zeta^{fric} \dot{x}^A_a \tag{1.66}$$

$$\dot{\zeta}^{fric} = \frac{\dot{f}}{Q} = \frac{\sum_j \frac{(p'_j)^2}{(m_j)s^2} - \psi k_B T}{Q} = \frac{3kN}{2Q} \left[T - \frac{2\psi}{3N} T_0 \right] \tag{1.67}$$

In the last step, the kinetic temperature is introduced; N is the number of particles, Q and ψ are parameters that influence the coupling of the thermostat and k_B is the Boltzmann

---
^{i} i going from 1 to 3N, a going from 1 to 3, A going from 1 to N

constant.[i] The derivative of the friction coefficient is proportional to the difference between the instant kinetic temperature and the chosen temperature $\frac{2\psi}{3N}T_0$. If the difference is big, the latter terms in Eq. 1.64 have a major influence on the dynamic of the system.

Andersen Barostat

The description of an isobaric process need the definition of a barostat. The Andersen Barostat[47] introduces the volume (V) depending coordinate ($x_a^A = V^{1/3}r_i$), see Eq. 1.63, and the corresponding momentum in the Hamiltonian, which was extended in order to describe the reduction of the volume due to a piston (with mass M and pressure P_0).

$$H = H_{Anders} = \sum_i \frac{1}{2m_i V^{2/3}} y_i^2 + \sum_{i\{j\}} \Phi(V^{1/3}r_i, V^{1/3}r_j) + \frac{1}{2M} Y^2 + P_0 V \quad (1.68)$$

The scaled system is defined in the coordinates: r_i, V and the corresponding momenta y_i, Y. The Hamilton formalism leads to the following formula:

$$\frac{dr_i}{dt} = \frac{dH}{dy_i} = \frac{y_i}{mV^{2/3}} \quad (1.69)$$

$$\frac{dy_i}{dt} = -\frac{dH}{dr_i} = -V^{1/3} \sum_{\{j\}} \frac{d\Phi(\{V^{1/3}r_i\})}{dr_i}$$

$$\frac{dV}{dt} = \frac{dH}{dY} = \frac{Y}{M}$$

$$\frac{dY}{dt} = -\frac{dH}{dV} = P_{ij} - P_0$$

In the previous equation, the pressure tensor is introduced, as follows from eq. 1.63. The equations of motion from the Hamilton Lagrange formalism are the following:

$$mV\ddot{r}_i = -\sum_j \frac{d\Phi(V^{1/3}r_i - V^{1/3}r_j)}{dr_i} - \frac{2}{3}m_i \dot{r}_i \dot{V} \quad (1.70)$$

$$M\ddot{V} = -(3V)^{-1}\left(-2(2mV^{2/3})^{-1}\sum_i y_i^2 + V^{1/3}\sum_i r_i \frac{d\Phi(\{V^{1/3}r_i\})}{dr_i} + 3P_0 V\right)$$

$$= \left(\sum_i P_{ii} - P_0\right) \quad (1.71)$$

In case of a strong deviation of the pressure P_{ij} from the reference value P_0, the changes in volume V will be significant. Eq. 1.71 contains the first time derivative of the volume which has the function of a friction term. Eq. 1.71 describes the dependance of pressure deviation on the fictitious mass and the second time derivative of the volume. The Andersen Hamiltonian leads to a NPE ensemble, with an equilibrated pressure P_0.

Parrinello-Rahman Barostat

As a generalisation to that, the constant pressure in the Parrinello-Rahman Barostat[48] is defined as a hydrostatic pressure. The Hamiltonian has the following functional form in terms

---

[i] In Gromacs MD-program a time constant τ is introduced instead of the parameters Q and ψ, it is chosen to be about 0.1 ps. $\dot{\zeta}_{gromacs} = \frac{4\pi^2}{\tau^2 T_0}[T - T_0]$

1.2. CLASSICAL MOLECULAR DYNAMICS SIMULATIONS

of screw-symmetric periodic (simulation) box. The scalar product between two vectors $\vec{v}$ and $\vec{w}$ is defined by $\vec{v} \cdot \vec{w} = \sum_{abc} v_a g_{ac} g_{cb} w_b$ in the screw symmetric vector space with respect to the screw symmetric basis vector (with components g_{ab}). The volume V in the screw symmetric system depends on the matrix of transformation g_{ab}, as already defined above. Now $\vec{x}^A = \sum_a x'^A_a g_{ab}$ is devided according to the screw symmetric vector space.

$$H = H_{(P-R)} = \frac{1}{2} \sum_{Acab} m^A \, x'^A_a g_{ac} g_{cb} x'^A_b + \sum_A \sum_{B \neq A} \Phi(\{((x'^B_a - x'^A_a) g_{ac} g_{cb} (x'^B_b - x'^A_b)\})$$

$$+ \frac{W}{2} \sum_{ab} \dot{g}_{ab} \dot{g}_{ba} + P_0 \sum_{abc} g_{1a} g_{2b} g_{3c} \epsilon_{abc} \quad (1.72)$$

Here Φ is a potential depending on the distance between two atoms A and B. Analogous to the approach above, one obtains the equation of motion for the screw symmetric case.

$$\frac{1}{2} m^A \sum_{bc} (g_{ac} g_{cb} + g_{bc} g_{ca}) \ddot{x}^A_b =$$ (1.73)

$$= - \sum_{B \neq A} \frac{d\Phi}{dx'^B_a} - \frac{1}{2} \sum_{Abc} m^A (\dot{g}_{ac} g_{cb} + \dot{g}_{bc} g_{ca} + g_{ac} \dot{g}_{cb} + g_{bc} \dot{g}_{ca}) \dot{x}^A_b$$

$$W \ddot{g}_{ij} = \sum_{Ak} m^A \, \dot{x}^A_i g_{jk} \, \dot{x}^A_k - \frac{d\Phi(g,x)}{dg_{ij}} - P_0 \frac{dV}{dg_{ij}} \quad (1.74)$$

$$= V g_{ij}^{-1} (3 P_{ij} - P_0)$$

The equations of motions are expressed in terms of the 3N coordinates x'^A_a and the 9 box vector components g_{ab}. The second derivative of the box vector components are proportional to the difference in the pressure tensor as defined by the pressure tensor from the virial definition by Park as explained above.[i]

Berendsen Thermostat and Barostat

The very stable Berendsen Thermostat and Barostat is obtained by a weak coupling to an external heat and/or pressure bath using the principle of least pertubation consistent with the required global coupling[49]. The modified equation of motion of the particles are given by Eq. 1.75 depending on the friction coefficient γ or time constant τ_T and the equilibrium temperature T_0.

$$m_i \dot{v}_i = F_i + m_i \gamma (\frac{T_0}{T} - 1) v_i = F_i + \frac{m_i}{2 \tau_T} (\frac{T_0}{T} - 1) v_i \quad (1.75)$$

The Eq. 1.76 introduces an equation of motion for the box vectors x depending on the constant pressure P_0 and the time constant τ_p and the isothermal compressibility β.

$$\dot{x} = v - \frac{\beta}{3 \tau_P} (P_0 - P) x \quad (1.76)$$

The equations of motions lead to a proportional scaling of velocity and the box vectors by the factor λ_T and λ_P. The factor from the termostat ($v = v \lambda_T$) equals [$\lambda_T = 1 + \frac{\Delta t}{2 \tau_T} (\frac{T_0}{T} - 1) \approx (1 + \frac{\Delta t}{\tau_T} (\frac{T_0}{T} - 1))^{\frac{1}{2}}$] and for the barostat ($x = x \lambda_P$) equals [$\lambda_P = 1 - \frac{\beta \Delta t}{3 \tau_P} (P_0 - $

---

[i] In Gromacs MD-Program [41] a time constant τ and the compressibility β_{ij} is introduced instead of the parameter W. $(W^{-1})_{ij} = \frac{4 \pi \beta_{ij}}{3 \tau^2 L}$; L is the longest box vector.

$P) \approx \left(1 - \frac{\beta \Delta t}{\tau_P}(P_0 - P)\right)^{\frac{1}{3}}$]. As a material depending parameter the compressibility β is introduced. The Berendsen algorithem are very stable and the equilibrium conditions are reached in only a short time[49]. The method does not alter drastically the dynamic properties of the system.[14]

1.3 Chemical Reactions

Chemical reactions transforming the system from reactant to the product state corresponds to the path from one local minimum to another crossing the saddle point on an energy hypersurface. The reaction is described by one or several parameters, the reaction coordinate ζ, which enables observation of any physical or chemical parameter during the reaction as an average of statistical thermodynamics depending on the chosen ensemble. The dimensionality of the reaction coordinate ζ depends on the complexity of the reaction.

1.3.1 Activation Energy

According to transition state theory, the reaction rates k of a chemical reaction are dominated by the activation energy E_a that is needed to bring the system to an activated state. An exponential dependence of the reaction rate on the activation energy was established empirically by the Arrhenius equation, see Eq.1.77. The concept of a activated complex is defined as the saddle point of the reaction.

$$k = Ae^{-\beta E_a} \tag{1.77}$$

Activation energies are obtained from the Arrhenius plot. Considering a physical property D which is dominated by only one process A at both temperatures T_1 and T_2, the activation barrier of the process E_A can be estimated from the Arrhenius Eq. 1.77 as follows.

$$E_A = (\ln[D(T_1)] - \ln[D(T_2)]) \frac{k_B}{1/T_1 - 1/T_2} \tag{1.78}$$

Error propagation leads to an error of the activation barrier of the process ΔE_A depending on the uncertainties in the property ΔD:

$$\Delta E_A = \left[\frac{\Delta D(T_1)}{D(T_1)} + \frac{\Delta D(T_2)}{D(T_2)}\right] \frac{k_B}{1/T_1 - 1/T_2} \tag{1.79}$$

In the micro canonical approximation the activation energy is given by the potential energy, albeit reaction conditions do not accomplish the conditions of an isolated system. In complex systems entropy has a major influence on the reaction kinetics due to conformational changes and density fluctuations. Then isothermal instead of adiabatic conditions are considered. Nevertheless, equilibrium conditions between activated complexes and the reactants are assumed and the motion of the system along the reaction path (at the activated state) is described as a free translation. This leads to the dependence of the reaction rate on the free energy instead of the potential energy.

1.3.2 Umbrella Sampling

The free energy $F(N, V, T)$ is defined by means of statistical thermodynamics as a function of the canonical partition function Z, $F = -\beta ln(Z)$, where $\beta = (k_B T)^{-1}$ is the inverse temperature (reciprocal of the temperature multiplied by the Boltzmann factor k_B). The

1.3. CHEMICAL REACTIONS

canonical probability distribution of states ρ with internal energy E is written as follows with Ω being the density of states.

$$\rho(E; N, V, T) = \frac{e^{-\beta E}\Omega(N, V, E)}{Z(N, V, T)} \tag{1.80}$$

$$\rho(\zeta; N, V, T) = \int \delta(E' - E(\zeta'))\delta(\zeta - \zeta')\Omega(N, V, E')\frac{e^{-\beta E'}}{Z(N, V, T)} \tag{1.81}$$

$$Z_\zeta = \rho(\zeta; N, V, T) \tag{1.82}$$

$$\Delta F(\zeta) = -\beta ln(Z_\zeta) \tag{1.83}$$

To obtain the free energy differences $\Delta F(\zeta)$ (Eq. 1.83) along a reaction coordinate ζ, a good statistics for the entire parameter interval is needed.

The sampling of rare reactions by normal MD simulation is not sufficient (in a certain simulation time) to calculate the free energy differences $F(\zeta)$ along the reaction path, because parts of the interesting interval of the reaction coordinate might not be sampled at all, as shown in part A of Fig. 1.2.

Therefore, the umbrella sampling is used, where an additional harmonic potential (bias) depending on the spring constant $k_{umbrella}$ and the equilibrium constant ζ_0 is introduced in the MD simulation that leads to a changed potential energy E'.[i]

$$E' = E + \frac{1}{2}k_{umbrella}(\zeta - \zeta_0)^2 \tag{1.84}$$

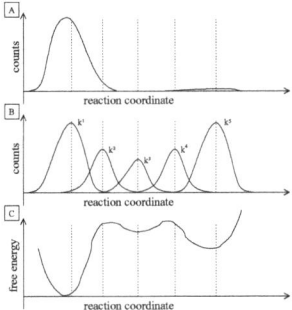

A strong spring constant leads to an increase of values of the reaction coordinate close to the equilibrium constant ζ_0 and a shift of the equilibrium constant ζ_0 to a certain value enables the sampling of a certain parameter range.

Several independent simulations are performed with different $\zeta_0 = \zeta_i$ and adequate $k_{umbrella} = k^i$. The interval $[\zeta_i - \Delta\zeta; \zeta_i + \Delta\zeta]$ covered by such an MD run is called sampling window with the distribution of the reaction coordinate $\rho_i^0(\zeta)$. The sampling is performed in such a way that the windows cover the whole parameter range and neighbouring ones overlap, as shown in part B of Fig. 1.2.

The biased probability distribution of states of the j-th simulation $\rho_j(\zeta)$ is related with the unbiased probability distribution of states $\rho^*(\zeta)$ through the exponential bias factor $c_j(\zeta) = e^{-\beta\frac{1}{2}k_j(\zeta - \zeta_j^0)^2}$ and the normalisation factor w_i.[ii]

Figure 1.2: A - sampling by conventional MD; B - sampling by several umbrella sampling runs with parameter ζ_i^0 and k_i; C - free energy surface

$$\rho_j(\zeta) = w_j c_j(\zeta)\rho^*(\zeta) \tag{1.85}$$

$$w_j = \frac{1}{\int d\zeta\, c_j(\zeta)\rho^*(\zeta)} \tag{1.86}$$

---

[i]Functional form as defined in the Gromacs manual

[ii]$\rho_x = \rho(\zeta)$ probability distribution of states with reaction coordinate $\zeta = x$, the continuous formulation can easily be transformed into a discrete one going to ρ_x from $\rho(\zeta)$ and to summation from integral

1.3.3 The Weighted Histogram Analysis Method

The weighted histogram analysis method (WHAM)[50] [51] seeks a linear combination $\tilde{\rho}^\star(\zeta)$ of M independent estimates of the probability distribution of states obtained from the measured biased distribution $\rho_j(\zeta)$, see Eq. 1.85, such that the integral over the variance ($\int d\zeta\, \sigma^2\,[\rho(\zeta)]$) is minimised. The contribution of each biased MD run j to a re-weighting estimate $\tilde{\rho}^\star(\zeta)$ is weighted by $\tilde{w}_i^{-1}$ based on the magnitude of errors in their histograms.

$$\tilde{\rho}^\star(\zeta) = \sum_i^M \tilde{w}_i^{-1} c_i^{-1}(\zeta) \rho_i(\zeta) \tag{1.87}$$

The conditional probability $P(\rho_j|\tilde{\rho}^\star, w_j)$ is the probability to obtain the distribution ρ_j under the condition that the unbiased distribution $\tilde{\rho}^\star$ and the weighting factor w_j are applied. Considering a random sampling from identical independent distributions, the probability $P(\rho_j|\tilde{\rho}^\star, w_j)$ factorises in the following way.

$$P(\rho_{jk}|\tilde{\rho}_k^\star, w_j) = \frac{gN_j!}{\Pi_k(gn_{jk})!} \Pi_k (\rho_{jk})^{gn_{jk}} \tag{1.88}$$

In order to find the best estimate $\tilde{\rho}^\star(\zeta)$ for the unbiased distribution $\rho^\star(\zeta)$, the conditional probability $P(\rho_{jk}|\tilde{\rho}_k^\star, w_j)$ is maximised under the condition of normalisation for each run j and value of ζ, referred to by index k. Therefore, the (logarithmic) maximum likelihood function L is defined, which is a statistical method used for fitting data to a statistical model[52, 30].

$$L = \sum_k \sum_j \ln P(\rho_{jk}|\tilde{\rho}_k^\star, w_j) + \sum_j \lambda_j \left(\sum_k \rho_{jk} - 1 \right) \tag{1.89}$$

The Lagrange multipliers λ_j are introduced to maintain normalisation independently. The derivatives of L are calculated and set to zero to find the maximum of the function L.

$$\frac{dL}{d\tilde{\rho}_k^\star} = \sum_j \frac{gn_{jk}}{\rho_{jk}} (w_j c_{jk}) + \sum_j \lambda_j w_j c_{jk} = 0 \tag{1.90}$$

$$\frac{dL}{dw_j} = \sum_k \frac{gn_{jk}}{\rho_{jk}} (c_{jk} \rho_k^\star) + \lambda_j \sum_k c_{jk} \rho_k^\star = 0 \tag{1.91}$$

here the number of counts in the j-th run is indicated as $N_j = \sum_K n_{jK}$. Combining the two conditions, Eq. 1.90 and Eq. 1.91, one obtains the WHAM equations:

$$\frac{\sum_i gn_{ik}}{\rho_{lk}^\star} - \sum_j c_{jk} \sum_K gn_{jK} w_j = 0 \qquad 0 = \sum_K gn_{jK} + \lambda_j \sum_K c_{jK} \rho_K^\star w_j$$

$$\sum_j gn_{jk} - \sum_j gN_j w_j c_{jk} \rho_k^\star = 0 \qquad 0 = \sum_K gn_{jK} + \lambda_j$$

In summary, the following formulae are usually referred to as WHAM method in the discrete (k) and continuous formulation (ζ):

$$\underbrace{\begin{aligned} \frac{N_x}{\sum_j N_j w_j c_{jx}} &= \tilde{\rho}_x^\star \\ w_j c_{jx} \tilde{\rho}_x^\star &= \rho_{jx} \\ \frac{1}{\sum_x c_{jx} \tilde{\rho}_x^\star} &= w_j \end{aligned}}_{\text{[discret]}} \Leftrightarrow \underbrace{\begin{aligned} \tilde{\rho}^\star(\zeta) &= \frac{N(\zeta)}{\sum_j N_j w_j c_j(\zeta)} \\ \rho_j(\zeta) &= w_j c_j(\zeta) \tilde{\rho}^\star(\zeta) \\ w_j &= \frac{1}{\int d\zeta\, c_j(\zeta) \tilde{\rho}^\star(\zeta)} \end{aligned}}_{\text{[continuous]}} \tag{1.92}$$

1.4 The Proton Coordinate

In quantum mechanical calculations, the proton is no distinguished particle as for example in the FF simulations. In FF description, the properties of all particles are predefined in the topology, whereas QM simulations allow general changes in the electronic structure, formation of bonds and displacement of the partial charges. The excess proton is assigned by the excess charge located in the proximity of a hydrogen atom. As an isolated proton does not exist in the chemical environment and stuctural diffusion is an important concept to describe proton transport e.g. in water, the proton coordinate is assigned as a physical property to the entire system, see Sec. 2.1.

The *modified centre of excess charge coordinate* (mCEC) defines the proton position by structural criteria[53]. It assigns "fixed partial charges"- weighting factors (w^A) - to each species ($A = \{X, H\}$) with coordinates $\vec{R}^A$ taking part in the structural definition of the proton, e.g. proton acceptor atoms (X) and all hydrogen covalently bonded to these atoms (H). The weight assigned to the hydrogen equals ($w^H = +1$), while the weight assigned to the proton acceptor molecule depends on the number of hydrogen atoms bound to them at the non protonated state, i.e. water oxygen has the weight (-2) and the nitrogen in imidazole has the weight (-0.5), see Eq. 1.97. The out-coming dipole moment is then corrected by further criteria to ensure a smooth and symmetric movement of the proton, see Eq. 1.93.

$$\vec{c} = \mathrm{mCEC} = \underbrace{\sum_A^N w^A \vec{R}^A}_{\text{dipole moment}} + \underbrace{(\vec{R}^X - \vec{R}^H)\vartheta^{XH}}_{\text{binding correction}} \quad (1.93)$$

$$+ \underbrace{\left[(w^{X'}\vartheta_{max}^{X'H'} - w^{X''}\vartheta_{max}^{X''H''})(\vec{R}^{X'} - \vec{R}^{X''})\right]_{\forall X', X'' \in (\text{same molecule})}}_{\text{geometrical correction}}$$

In Eq. 1.93, the binding correction depends on the switching function ϑ, which is one for short distances between the atoms of type H and X, and zero for long distances, see Eq. 1.94. The characteristic of ϑ is influenced by two parameters k_0 and k_1, which influence the position of the inflection point and gradient at that point, see Fig. 1.3. $\vartheta(x)$ equals 1 if the distance x is smaller than the chemical bond distance of about 0.1 nm and 0 if the distance matches the distance of hydrogen bond, about 0.18 nm. In between these points the function has a smooth transition.

$$\vartheta^{XH} = \vartheta(\vec{R}^X - \vec{R}^H) \quad (1.94)$$
$$= \left(1 + \exp(k_1 \left|\vec{R}^X - \vec{R}^H\right| - k_0 k_1)\right)^{-1}$$

For the geometrical correction in Eq. 1.93, X' and X'' are defined as non-hydrogen atoms (X) of the same molecules. The term $\vartheta_{max(H)}^{X'H'}$ refers to the maximal value ϑ^{XH} belonging to X', see Eq. 1.95. This term plays a role in molecules which contain several undistinguishable atoms X, with not all of them being bound to exactly the same number of hydrogen atom in the neutral state. An example of such molecules is imidazole, which has two equivalent nitrogen atoms with only one of them bound to a hydrogen.

$$\vartheta_{max}^{X'H'} = \max_{H' \in H} \vartheta^{X'H} \quad (1.95)$$

Considering the following parameters, $k^0 = 0.12$ nm and $k^1 = 300$ nm^{-1},[i] the transition

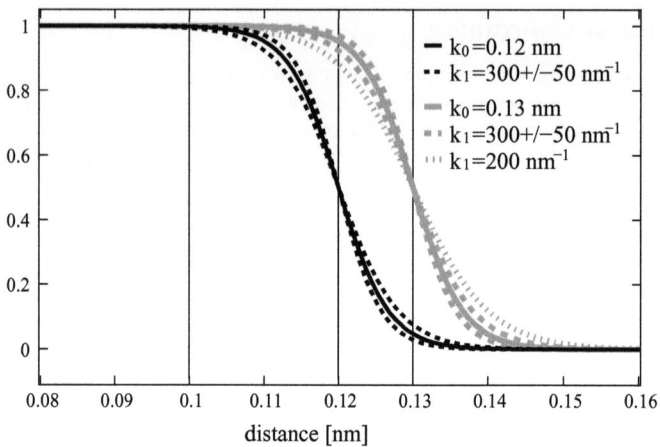

Figure 1.3: *The switching function ϑ depending on the parameters k^0 and k^1*

area is about 40% of the bond length. The choice of the parameters will have an influence on the mCEC and also on the reaction coordinates. On the one hand, a broad transition region favours the description of a smooth reaction coordinate, on the other hand, it leads to wrong location of the proton, as the description of the chemical bond is not sufficiently fulfilled.

In order to implement the coordinate in a MD program derivatives were calculated that are added in the Appendix B.3.1.

In the case of imidazole molecules, Formula 1.93 turns into th following.

$$\vec{c} = \mathrm{mCEC} = \underbrace{\sum_{i=1}^{N_H} \vec{R}^{H_i} - \sum_{j=1}^{N_N} 0.5\vec{R}^{N_j}}_{\text{dipole moment}} \tag{1.96}$$

$$- \underbrace{\sum_{i=1}^{N_H}\sum_{j=1}^{N_N} \theta\left(R^b - \|\vec{R}^{H_i} - \vec{R}^{N_j}\|\right) \cdot \left(\vec{R}^{H_i} - \vec{R}^{N_j}\right)}_{\text{binding correction}}$$

$$+ \underbrace{\sum_{J=i}^{N_{imidazole}} 0.5\left(\vec{R}^{(N\cdot)_J} - \vec{R}^{(NH)_J}\right)}_{\text{geometrical correction}}$$

As mentioned before, the first two terms of Eq.(1.97) correspond to the total dipole moment, which consists of a summation over all position vectors of the total number of hydrogen atoms (N_H), that are bound to nitrogen atoms weighted with an approximate charge factor of $+1$ and respectively all the position vectors of the total number of nitrogen atoms (N_N) with a factor of -0.5. The function $\theta\left(R^b - x\right)$ in the third term is a switching function depending

---

[i]The inflection point at $k^0 = 0.13$ nm equals the distance of intermolecular hydrogen in protonated zundel-like imidazole complexes as explained later in section 3.1.

1.4. THE PROTON COORDINATE

on the maximal bond length between hydrogen and nitrogen ($R^b = 1.3$ Å). This term is a geometrical correction that cancels the bond vector in case of covalent bound hydrogen to a nitrogen, or hydrogen to a oxygen respectively for water. The fourth term describes an intramolecular correction, therefore the summation is taken over all imidazole molecules, $N_{imidazole}$ refering to their total number. This term is only useful in case of molecules with more than one proton acceptor atoms, among which at least one is covalently bound to a hydrogen in the non protonated state. Here ($N\cdot$) and (NH) mean the deprotonated and protonated nitrogen of a molecule[54].

Chapter 2

Proton Transport and PEM Additives

This chapter aims to narrow the broad research field concerning PEM fuel cells towards the issue of proton transport in the PEM additives. In Sec. 2.1, the general concept of proton transport, especially in the context of water based transport, is introduced and the proton conducting species are presented. Subsequently, the PEM additives as characterised by experiments in the literature are described in detail and an overview is given with respect to their performance as additives in PEM, see Sec. 2.2.

2.1 Protonation of Molecules and Proton Transport

In standard PEM material, proton transport often relies on water. Therefore, proton transport in water is briefly discussed, followed by the presentation of two other important molecules for proton transport, sulphonic acid and imidazole, and their protonation state.

2.1.1 Proton Transport in Water

Two borderline mechanisms contribute to proton transport in water, one being the vehicle mechanism which is oriented at the transport of other ionic species such as Na^+ or Cl^- and involves the identification of a ionic particle in water.

The definitions of the Eigen cation ($H_3O^+\times(H_2O)_3$; 1954 [55, 56]) and the Zundel cation ($H_5O_2^+$; 1968[57]) mark the path towards the concept of structural transport, which was already introduced by Grotthus in 1806[58]. The proton solvation and transport in water have been studied by vibrational and rotational spectroscopic methods[59], which determined various ionic water complexes. Therefore, the proton cannot be identified as a specific hydrogen atom or fixed group of atoms. Technically, this issue is reflected in the need of a proton coordinate in (QM based) computer simulations as introduced in Sec. 1.4. The structural proton diffusion involves continual interconversion of covalent and hydrogen bonds, which results in a dynamic equilibrium of Eigen and Zundel ions and even larger protonated water clusters[60, 61, 62]. The rate limiting step is the fluctuation induced breakage of hydrogen bonds in the first and second solvation shell[61, 62]. The formation of a hydrogen bond network leads to a strong correlation of the dynamic in the system[63, 64]. The strength of hydrogen bonds in water is about 2 kcal/mol [64] and their life time is in the range of picoseconds[60, 64]. It has been shown by Tuckerman et al.[65] that the atoms in a Zundel cation behave in an essentially classical manner at room temperature and an adiabatic approximation can be

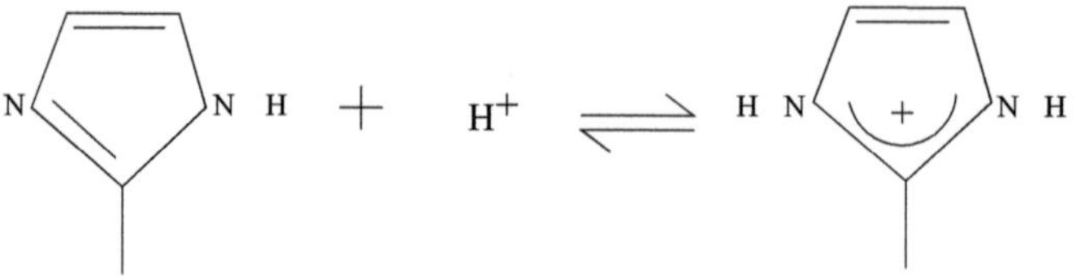

Figure 2.1: *Scheme of protonation of neutral imidazole. The proton and hydrogen bonded to nitrogen are explicitly shown.*

applied to the proton transport[65].

Proton diffusion in water is determined by experiment ($0.93 \cdot 10^{-8} \text{m}^2\text{s}^{-1}$)[66] and it was shown that the structural diffusion mechanism dominates for low temperature, whereas the vehicle mechanism takes over in liquid water at temperatures above 430 K[64]. Proton diffusion has also been the subject of various simulational studies. The published ab initio simulations tend to underestimate the diffusion coefficient by a factor of three to five[67, 68, 69].

2.1.2 Other Proton Transport Species

Sulphonic acid is already known from the early days of the fuel cell and nowadays the species plays an important role as the hydrophilic functional part of the most commonly used PEM, Nafion. Furthermore, proton transport and protonic states are important issues in biological systems where imidazole is well known as a component of the amino acid histidine, as well as sulphonic acid is a part of the amino acid taurin. Besides the amphoteric character and the possibility of autodissociation, extraordinary thermal and chemical stability makes both species, imidazole and sulphonic acid, interesting candidates also for technical applications. In Nafion, autodissociation of the acid leads to a high concentration of intrinsic charge carriers under hydration. Since the autodissociation in imidazole is quite low, the intrinsic charge carrier concentration in imidazole is not sufficiently high and such system is usually doped with another material. For proton conductivity owing to a fluctuating hydrogen bond network, the rate of fluctuation has a stronger influence on the conductivity than the concentration of charge carriers[70]. For mixed systems of imidazole and Nafion, an adequate conductivity was measured especially at high temperatures [70].

Imidazole

Imidazole is a neutral nitrogen containing carbon 5 ring, which has two inequivalent nitrogen atoms, since only one is bound to a hydrogen atom. Protonation of the second nitrogen atom leads to an aromatic system, see Fig. 2.2. The two equivalent hydrogen atoms in the protonated imidazole molecule enable proton transport along the hydrogen bonds towards a neighbouring molecule, similar to structural diffusion.

In imidazole crystals, two different pathways have been proposed, both involving reorganisation of hydrogen bonds as the rate limiting step. The dominating pathway could not be satisfactorily clarified by NMR analysis[71, 72]. Unlike the pathway involving charge separation, a Grotthus like character of the pathwah was confirmed by ab initio Car-Parrinello MD[73]. Proton hopping between two molecules happens in about 0.3 ps and the reorientation is estimated from the diffusion coefficient ($2 \cdot 10^{-9}$ m^{2}s^{-1} at about 400 K), which results in a time step of 30 ps[74]. The activation energy for proton diffusion is about 40 kcal/mol[75].

2.1. PROTONATION OF MOLECULES AND PROTON TRANSPORT

In ab initio computer simulations, Zundel like imidazole configurations are observed with an intramolecular nitrogen to nitrogen distance of 0.281 nm[76, 73].

In liquid imidazole, proton diffusion similar to liquid water consists of both mechanisms: the hopping of a proton from one molecule to another, complemented with structural diffusion and the continuous movement of protonated molecules, called vehicular diffusion. In the literature, the vehicular diffusion contributes to one third of the diffusion in liquid fluorinated methyl imidazole at a temperature of 450 K [i]. At this temperature, the dominance of the structural mechanism in imidazole is in parallel to the case of liquid water with respect to the boiling and melting point.

As already mentioned above, autodissociation of imidazole is quite low. Being a base, the protonation of imidazole is favoured in aqueous solution, as confirmed by potentiometry. The free energy difference between the two states is estimated to be about 9 kcal/mol[77]. For histidine[ii], deprotonation is studied in the presence of 50 water molecules using thermodynamic integration (using Car-Parrinello MD with a BLYP functional)[78, 79] and for a full solvation model using umbrella sampling (using the empirical valence bond method)[80]. These studies report a barrier for deprotonation of 8 to 9 kcal/mol in water environment.

Sulphonic Acid

Sulphonic acid is an organo sulphur compound with the formula R-S(-OH)(=O)$_2$. Its protonation involves the formation of a second covalent bond between oxygen and hydrogen atoms, while bond breaking leads to deprotonation that results in the negatively charged group, R-SO$_3^-$. The main advantage of sulphonic acid based systems is the existence of intrinsic charge carriers, as deprotonation of the acid occurs even under reduced hydration. Even under low hydration, transport rather of a deficit than of an excess proton is expected in the sulphonic acid system.

From measurements such as infra red spectroscopy, Raman spectroscopy and nuclear magnetic resonance spectroscopy, the anionic concentrations in aqueous solutions in the case of methyl sulphonic are known[81, 82, 83]. At room temperature, the highest concentration is reached for a 2:1 ratio of water molecules to methyl sulphonic acid. The relative deprotonation of methyl sulphonic acid groups increases with water content, 90 percent is reached for a 7:1 ratio and at a ratio of 11 water molecules per acid more than 95 percent are deprotonated.

For methyl sulphonic acid, deprotonation is studied in the presence of small water clusters by ab initio methods[84, 85]. Deprotonation is predicted to occure in the presence of more than 3 water molecules. The analysis is based on potential energy and frequency calculation to estimate the free energy. In the case of trifluronated methyl sulphonic acid, which is used as a model system for Nafion, similar results are obtained by ab initio method[86]. Deprotonation takes place at a cluster size of more than 3 water molecules and the authors state complete deprotonation, i.e. the detachment of the proton from the anion, at more than 6 molecules. The first conclusion has been confirmed by the so called metadynamics calculation of free energy[87]. Furthermore, a DFT MD simulation has been performed for a Nafion model; the obtained free energy barriers were in the range of about 2 kcal/mol with a water content of up to 25 water molecules and two sulphonic acid groups of Nafion at 360 K.[88].

Accordingly, the proton transport is strongly based on diffusion in water. In Nafion, proton transport exclusively occurs in the hydrophilic part of the phase separated polymer.

---

[i]$T = 450K$: $D_V = 0.15 \frac{10^{-9} m2}{s}$, $D_H = 0.3 \frac{10^{-9} m2}{s}$ [67]
[ii]Histidine is an amino acid containing an imidazole group, as mentioned above.

Eventhough, the microscopic structure is still under discussion, a strong influence of hydration is clearly observed[8, 89]. Upon hydration, nanometer-sized water channels are formed by sulphonic acid containing side chains. These hydrophilic regions are surrounded by the hydrophobic carbon-fluoro backbone. At lower water content, the continuous water network is disconnected by hydrophobic regions leading to a breakdown in proton conductivity[8, 9].

According to experimental studies, the diffusion coefficient of Nafion is in the range of 0.5-2 $\cdot 10^{-5} cm^2 s^{-1}$ and activation energies are in the range of 3 to 5 kcal/mol[89, 90, 91, 92, 93]. The activation energies from MD simulations are reported to be 4 kcal/mol for water and about 2 kcal/mol for the hydronium ion at different water contents[94, 95]. From classical MD simulations for a Nafion system with a water sulphonic acid ratio of $\lambda=14$ at 300 K, the diffusion coefficient for water equals 0.6 $\cdot 10^{-5} cm^2 s^{-1}$ while the diffusion coefficient for the hydronium ion is lower by a factor of 3 (0.2 $\cdot 10^{-5} cm^2 s^{-1}$)[94].

2.2 PEM Additives - Experimental Insight

The performance of conventional PEM material, such as Nafion or S-PEEK, strongly rely on temperature and pressure conditions, especially due to water dependant proton transport mechanisms. To overcome these limitations, additives to the PEM aim to compensate the failure at elevated temperatures and low hydration conditions. Since inorganic particles are known for thermal stability and for water retention at higher temperatures[11, 3], the task is tackled by composit material of thermally and structurally stabile inorganic compounds and highly conductive functional groups[96, 97, 98], to improve the intrinsic proton conductivity of the material. The following overview refers to the work of Marschall et al. as these additives have been in the focus of the computational research.

2.2.1 Characterisation of the Functionalised MCM-41

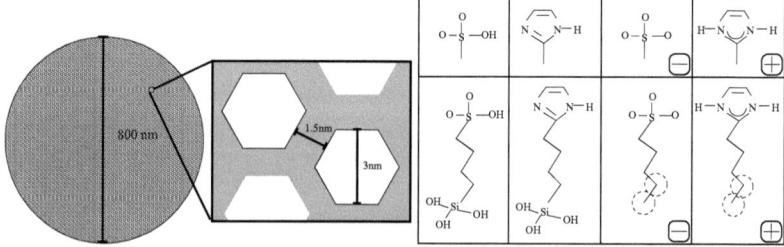

Figure 2.2: *Schematic figure:* **Left:** *MCM-41 particle;* **Middle:** *Close view of the pore, light grey amorphous silicon dioxide;* **Right:** *Molecules (left to right, first and second row): methyl sulphonic acid, methyl imidazole, deprotonated sulphonic acid and protonated methyl imidazole, trihydroxy silanyl propyl sulphonic acid, trihydroxy silanyl propyl imidazole, deprotonated pentyl sulphonic acid and protonated pentyl imidazole. The dotted lines mark carbon atoms of the terminal ethyl group, see Sec. 3.2.2. The circle denotes the total charge of the molecules.*

The *Mobile composite of matter 41* (MCM-41) was first described in 1992 by Beck et al.[99, 100] and is an amorphous silicon dioxide material with a density of about 2.2 g/cm^3 in the shape of highly ordered hexagonal nm sized pores, see Fig. 2.2. The surface density of hydroxyl groups was estimated to be about 3 to 8 nm^{-2}[101, 102]. To enhance proton conduc-

2.2. PEM ADDITIVES - EXPERIMENTAL INSIGHT

tivity for fuel cell application, the MCM-41 is modified by incorporation of organic molecules, i.e. sulphonic acid or imidazole containing molecules. The functional molecules are propyl sulphonic acid or propyl imidazole (see Fig. 2.2) which are covalently bound to the silanol material.

Two different synthesis routes exist[103]. The grafting route consists of two separated steps; in the first step, the MCM-41 material is built and, in the second, the material is functionalised via a condensation reaction. MCM-41 is instead built simultaneously with its functionalisation in the co-condensation synthesis route. In grafted particles the distribution of functional groups was shown to be more inhomogeneous because of pore blocking, which hinders the functional molecules to enter in the middle of the pore during functionalisation. For co-condensation, the limiting factor in functionalisation is the loss of the mesoporous structure by increasing the functional loadings[104, 103].

The μm sized additives have a rough surface[101, 105]. By reducing the size further to about

Table 2.1: *Characterisation of material with MCM-41 host: The degree of functionalisation is specified in [mmol g^{-1}] as chosen for the synthesis; Porosity is characterised by the inner surface, the pore volume and the pore diameter. There Parameters are obtained from nitrogen isotherms by BET and BJH method; The lattice constant of the ordered hexagonal mesoporous structure of the MCM-41 is measured by X-ray diffraction; Ion exchange capacity (IEC) are listed for the sulphonic acid containing system and proton acceptance capacity (PAC) in the case of imidazole. All data is taken from the literature, citations are denoted in the table.*

funct. [mmol g^{-1}]	surface [m^{2}g^{-1}]	volume [cm^{3}g^{-1}]	diameter [nm]	lattice [nm]	IEC/PAC [mmol g^{-1}]
grafted sulphonic functionalisation [101]					
0	1031	0.9537	2.7	4.53	-
5					
20					≈1.67
co-condensation of sulphonic functionalisation [103]					
0	1030	0.954	2.7		
10					
20	645				
30	625				
40	-		1.5		2.3
nanoparticles; co-condensation of sulphonic functionalisation [104]					
0	1459	1.34	2.8	4.7	
10	1418	1.31	2.5	4.55	0.84
20	1135	0.74	2.1	3.72	1.78
grafted imidazole functionalisation [105]					
	1181	1.128	2.7	4.53	
10	1126	0.948	2.1	4.53	0.21
20	953	0.825	2.1	4.53	0.48

0.08 μm, the properties of the additives could be improved and a better incorporation in the PEM is expected[104]. Proton conductivity is influenced by the functionalisation, as well as the degree of porosity, the size and the distribution of pores[106]. Therefore, the porosity was characterised by nitrogen adsorption measurements which give the pore volume, surface area and pore diameter. The lattice constants of the ordered porous structure is 3.7 nm to 4.7 nm for the samples and the wall thickness equals about 1.8 nm[101, 105], see the Schematic view in Fig. 2.2.

In Tab. 2.1, the properties of the different functionalised MCM-41 material are listed as reported by Marschall et.al.. The degree of functionalisation is given by the number of functional groups in the synthesis of one gramm. Furthermore, the listed ion exchange capacity (IEC) and proton acceptance capacity (PAC) are a measure for the number of functional groups. An estimate for the surface density of sulphonic acid groups results from dividing the IEC by the inner surface of the pore. The changes in pore diameter, volume and surface after functionalisation derive from changes in the surface interaction and roughness, as well as from the space occupied by the functional groups. For the co-condensation materials it was shown by calcination of the organic groups, that the mesoporous structure of MCM-41 remains unchainged for degrees of functionalisation up to 40 mmol g^{-1}. The distribution of functional molecules in the porous environment was clarified by neutron scattering diffraction[107]. The repression of matching with a D_2O/H_2O mixture that was chosen to match the unfunctionalised MCM-41 material, hints to a homogeneous distribution of functional groups in the material from co-condensation, while the matching in the case of grafted materials results from non functionalised regions inside the pores.

In summary, the additives are of μm size and have a highly ordered hexagonal mesoporous structure. The MCM-41 host material has a large inner surface of 1000 to 1500 m^{2}g^{-1} and a large pore volume of about 1 cm^{3}g^{-1}. The pore diameter is in the range of 3 nm and the pore wall is less than 1.8 nm thick. A homogeneous functionalisation of up to 1.4 groups per nm^2 is obtained via co-condensation of sulphonic acid functional groups[103]. In the case of imidazole only the grafting synthesis route was performed.

2.2.2 Proton Transport in the Material

Table 2.2: *Conductivity at different temperatures and humidities of functionalised MCM-41, as reported by Wark et.al.. Citations are denoted in the table.*

short description of the material	conductivity [Scm^{-1}]			
	T= 330 K		T= 410 K	
	50 RH	100 RH	50 RH	100 RH
20 mmol/g grafted; sulphonic [101]	10^{-8}	10^{-5}	10^{-4}	10^{-3}
20 mmol/g co-condensated; sulphonic[103]	-	10^{-5}	-	10^{-2}
40 mmol/g co-condensated; sulphonic[103]	10^{-4}	10^{-3}	10^{-2}	10^{-1}
20 mmol/g grafted; imidazole [105]	10^{-7}	10^{-6}	10^{-5}	10^{-4}
Nafion [103]	-	10^{-2}	-	10^{-4}

Proton transport in these materials is characterised by impedance spectroscopy of pellets. An overview of the conductivities for different humidity conditions, temperatures and materials is given in Tab. 2.2. The relative humidity varies between 100%-relative humidity and 50%-relative humidity. For all functionalised materials and temperatures, the conductivity increases with the humidity by at least one order of magnitude and by at least two orders of magnitude increasing the temperature from 330 K to 410 K. The strongest dependence on the water content is observed for the grafted sulphonic acid system at the lower temperature. The sulphonic acid based system also show the stronges dependence on temperature.

Even if only the sulphonic acid co-condensation material is competitive with Nafion, from the point of view of the conductivity, the material shows promising tends at higher temper-

2.2. PEM ADDITIVES - EXPERIMENTAL INSIGHT

atures and low humidity. Sulphonic acid functionalised mesoporous particles were added to different polymers, polysiloxanes as well as polyoxadiazoles[108, 109]. These hybrid materials showed improved proton conducting abilities, but so far no improvement of mechanical properties was reported.

Part II
PEM Additives - Computer Simulation

Chapter 3
Proton Transport Species

In the previous chapter, proton transport - being an important material property - was discussed for water, imidazole and sulphonic acid species and the studied class of materials was introduced. The silicon dioxide material is functionalised by organic molecules, alkyl imidazole and alkyl sulphonic acid, as shown in Fig. 2.2. This chapter describes calculations of characteristic physical properties for both functional groups as well as for water molecules neglecting the effect of the anorganic substrate.

In Sec. 3.1, simple properties of the proton conducting groups are calculated that first of all serve as a validation of the method DFTB with the *pbc/mio* SK-file set, but they provide at the same time important information about the proton conducting abilities of the groups. In the next section, the effect of immobilisation of the molecules via a carbon chain and the influence of their geometrical arrangement are studied. To this aim, proton diffusion coefficients are calculated in vacuum (Sec. 3.2.2) and hydrogen bond fluctuation and aggregation of the isolated systems (Sec. 3.2.3) are investigated.

3.1 Proton Conducting Species - Validation

Water molecules are linked by a hydrogen bond network which leads to the formation of water clusters and is also known to influence proton transport. A measure for the intermolecular energy is the binding energy of a cluster, which is calculated for water in Sec.3.1.2. The proton affinity depending on potential energy differences is a simple property to characterise the proton conducting species, water, (methyl) imidazole and (methyl) sulphonic acid, in Sec. 3.1.3. Besides the properties depending on an isolated molecule, proton transport is approached by potential energy barriers for proton transport between two groups, see Sec. 3.1.4. The model systems and the quantum mechanical methods are briefly described in Sec. 3.1.1.

3.1.1 Model Systems and QM Methods

In order to calculate potential energy differences for different configurations of the proton conducting groups, small molecular systems were considered neglecting the chemical environment. In this case, methyl groups are a good approximation for longer alkyl chains as the bonds in the carbon chain are known to be non-polar and the influence of the chain is therefore only small. Besides isolated methyl imidazole and methyl sulphonic acid, water clusters of a size (n) up to 20 molecules were described. The clusters of lowest energy have been taken from the Cambridge Cluster Database[110][111][112][113].

QM geometry optimisations and energy calculations were performed at three different levels of theory, see Sec. 1.1.2 and 1.1.3. Using the NWCHEM code[114], geometrically optimised energies based on a post Hartree-Fock method and on the DFT method were calculated. The post Hartree-Fock method of choice was the Møller-Plesset perturbation theory to the second order term, denoted MP2[22]. The DFT calculations were performed under using the hybrid exchange correlation functional B3LYP. For both ab initio methods, the correlation-consistent polarised valence basis sets, denoted cc-pVDZ, cc-pVTZ, cc-pVQZ, were considered in order to achieve convergence regarding the number of basis functions in the consecutive basis sets. The DFT calculations were also performed with the commonly used polarisable Popel basis set 6-31g*.

As the third QM method, the DFTB method has been applied. The calculations were run with the DFTB+ code[115]. The DFTB parameters[42], band structure integrals and two particle interaction parameters were taken from the SK-files registered in the following SK-file set: *mio-0-1* SK-file set[i] and *pbc-0-2* SK-file set[ii]. Since parameters from different SK-files are not optimised for combined use, the DFTB description was validated carefully for the actual systems comparing to the ab initio methods. In general, DFTB method is known to underestimate ordering of second or higher order water neighbours and the binding energy[116]. Therefore, Hu et al.[34] introduced additional (empirical) parameters to DFTB in order to improve the description of hydrogen bond interactions in water. This approach has been introduced in Sec. 1.1.3 as H-bond-DFTB and is used in the following in addition to the primal DFTB approach. For validation, all three levels of theory were considered for the BE and PA calculation, but the potential energy barrier was only calculated by DFTB and DFT.

3.1.2 Binding Energy

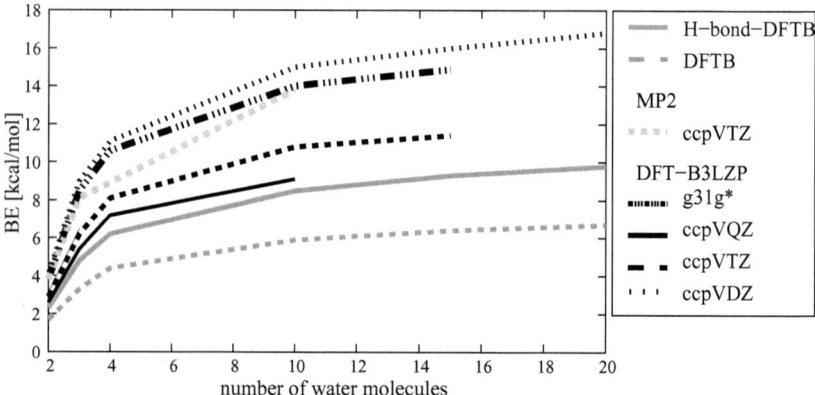

Figure 3.1: *Binding energies per molecule (BE) of water clusters depending on the cluster size n at different levels of theory: light grey lines - MP2 with cc-pVDZ basis set, black lines - DFT with cc-pV(D/T/Q)Z basis set or 6-31g* basis set; dotted and dashed lines refer to different functional basis set, see Table A.1; grey line - H-bond-DFTB, dashed DFTB.*

---

[i]Two particle interaction: C-S, S-S, S-O, H-H, H-O, S-H, N-N, N-H
[ii]Two particle interaction: C-C, C-H, C-O, C-N, N-O

3.1. PROTON CONDUCTING SPECIES - VALIDATION

The *binding energies per water molecule* $(BE)^i$ for the water clusters (nH_2O) are calculated, see Eq. 3.1.

$$BE = E_{(H_2O)} - \frac{1}{n}E_{(nH_2O)} \qquad (3.1)$$

The BE depends on the difference between the potential energy of an isolated water molecule $E_{(H_2O)}$ and the potential energy of an isolated water cluster $E_{(nH_2O)}$ divided by the number of molecules (n).

In Fig. 3.1, the BE of water clusters is shown, for the exact values see Table A.1 in the Appendix. The DFTB, H-bond-DFTB, DFT and MP2 results are compared to DFT data from the literature[111]. The results from DFT with B3LYP functional and 6-31g* basis set are in good agreement with results from the literature with comparable basis set and functional, DFT(B3LYP/6-31G(d,p)), not shown in Fig. 3.1, see Table A.1. The ab initio results show a strong dependence on the basis set. An enormous decrease in the binding energy per molecule appears as the cc-pV(D/T/Q)Z basis set is augmented. The convergence with respect to the basis set is hardly reached for cc-pVQZ.

As expected[116], the BE is underestimated for higher water clusters comparing primal DFTB to all other results. The BE calculated by H-bond-DFTB are increased by up to 3 kcal/mol for the large clusters compared to primal DFTB. The BE in the case of 2 water molecules is about 3 kcal/mol for DFT and MP2 with the cc-pVTZ basis set and about 1 kcal/mol lower for DFTB and H-bond-DFTB. Increasing the cluster size to 10 molecules, the BE is increased by 8 kcal/mol for DFT with the same basis set and by 10 kcal/mol for MP2 with the smaller basis set. In comparison, the DFTB and H-bond-DFTB method lead to an increase in BE of 4 kcal/mol and 6 kcal/mol, respectively. However, the DFTB and H-bond-DFTB method lead to reasonable results for the BE considering the high dependence of ab initio calculations on the basis set.

3.1.3 Proton Affinity

The *proton affinities* (PA) of methyl imidazole, methyl sulphonic acid and the water clusters are presented. The PA is defined as follows, here X stands for neutral methyl imidazole $(CH_3 - C_3N_2H_3)$, neutral methyl sulphonic acid $(CH_3 - SO_3H)$ and deprotonated/negative methyl sulphonic acid $(CH_3 - SO_3^-)$:

$$PA = E_{(nH_2O)} - E_{(nH_2O+H^+)} - E_{(H^+)} \qquad (3.2)$$
$$PA = E_{(X)} - E_{(X+H^+)} - E_{(H^+)} \qquad (3.3)$$

The terms $E_{(nH_2O)}$, $E_{(nH_2O+H^+)}$, $E_{(H^+)}$, $E_{(X)}$ and $E_{(X+H^+)}$ denote the potential energy of the corresponding isolated geometrically optimised system as referred to in brackets.

For the DFT method, the energy value $E_{(H^+)}$ (Eq. 3.2 and Eq. 3.3) are zero as the nuclear contribution is not taken into account[117]. For the DFTB method, $E_{(H^+)}$ corresponds instead to the repulsive energy of the proton and is set to $E_{(H^+)} = 142$ kcal/mol, which is obtained from substraction of the DFTB band structure term (Eq. 1.33) from the total energy of a neutral hydrogen atom. The latter was calculated by the DFT with B3LYP functional and (6-31++G**) basis set by Zhou et al.[117].

$$E_{DFT} - E_{BS} = 142 \text{ kcal/mol} \qquad (3.4)$$

---
[i]The binding energy is also called stabilisation energies of the cluster in the literature [111]

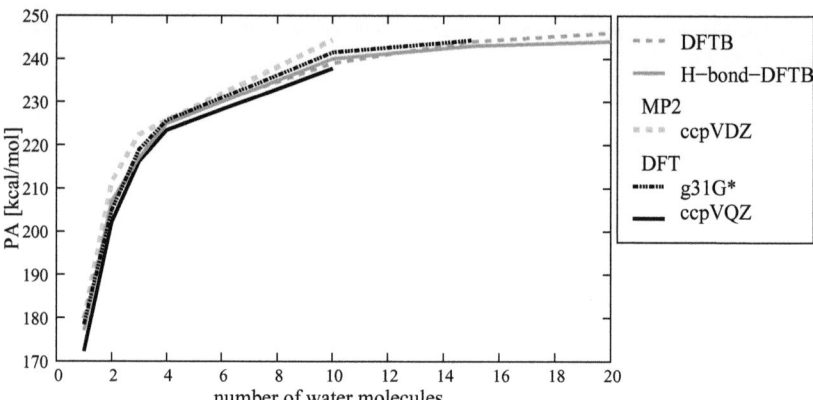

Figure 3.2: *Proton affinities (PA) of water clusters depending on the cluster size, see also Table A.2 at different levels of theory: light grey lines - MP2 with cc-pVDZ basis set, black lines - DFT with cc-pVQZ basis set or 6-31g* basis set; dotted and dashed lines refer to different functional basis set; grey line - H-bond-DFTB, dashed DFTB.*

This approach was introduced in order to minimise the deviation from ab initio PA calculations, because the repulsive energy for the bare proton was underestimated by about 10 kcal/mol taking directly the the repulsive energy from the hydrogen SK-file sets. The repulsive energy arises from fitting the energies of an entire parameter set, but the properties of the isolated proton are far different from a hydrogen atom in a molecular environment.

In Fig. 3.2, the PA of water is shown, for the complete list see Table A.2. Both DFTB methods are in good agreement compared to DFT with B3LYP functional and the largest basis set cc-pVQZ. The PA calculated by the primal DFTB method is overestimated by up to 4 kcal/mol for all water clusters with ($n \geq 2$), which is in the same range as going to a smaller basis sets, cc-pVTZ or 6-31g*. The H-bond-DFTB results deviate by only 1 kcal/mol from the results of primal DFTB. The values calculated through the MP2 method with small basis set are comparable to the DFT values (with the same basis set) for the water clusters. The PA of a single water molecule is overestimated by 5 kcal/mol and 8 kcal/mol comparing H-bond-DFTB results and primal DFTB results with DFT (B3LYP/cc-pVQZ), respectively. The proton affinity of imidazole is lower by about 6 kcal/mol, while sulphonic acid is even lower by only 7 kcal/mol than results from DFT(B3LYP/cc-pVQZ). In all cases, the relative error is below 3%.

As PA depends on the protonic repulsive energy $E_{(H^+)}$, energy differences between species and clusters are more important than the absolute differences between different methods. The PA of an isolated imidazole is for any method in the range of the PA for a water cluster of about 10 water molecules.The deviation of DFT with 6-31g* from all other results is big. The difference in PA of imidazole is about 15 kcal/mol comparing to the DFTB result and about 10 kcal/mol comparingn to the DFT methods with a different basis set. In general the imidazole PA matches the PA of the largest water cluster, we expect to obtain reasonable results by both DFTB methods for a system where the competing species consist of water cluster and imidazole. Due to unscreened electrostatic interaction, the proton affinity of deprotonated (negatively charged) sulphonic acid is about 100 kcal/mol higher than the

3.1. PROTON CONDUCTING SPECIES - VALIDATION

imidazole system.

3.1.4 Proton Transport Barrier

The approximated potential energy barrier for the proton transport between two proton conducting species is obtained from restrained energy minimisations in a system of either two methyl imidazole molecules, two deprotonated methyl sulphonic acid molecules or two water molecules. Each system contains an additional proton, which leads to a total of (-1) electronic charge in the case of the sulphonic acid system and a total of (+1) electronic charge for the other systems.

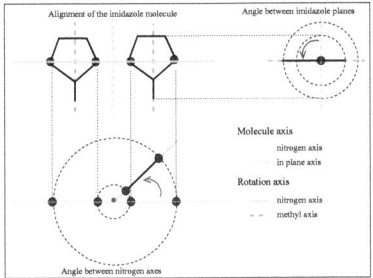

Figure 3.3: *Schematic figure: definition of the imidazole molecule axis.*

Computational Details

Besides the choice of proton conducting groups, a further parameter characterising the system is the distance between the groups, which is restrained by position restraints on either one nitrogen atom or one oxygen atom of each molecule, respectively. To avoid rotations of methyl imidazole and methyl sulphonic acid, the carbon atom of the methyl groups is restrained in addition. For each system with fixed distance of the proton conducting groups, the energy characteristics along the path of the proton between the groups is scanned stepwise by restrained energy minimisations, keeping not only the distance between the groups fixed, but also the position of the proton. The difference between energy minimum and energy maximum, i.e. the proton transport barrier, is given as a function of the distance between the species for each system.

In the case of methyl imidazole molecules, the influence of the angle between the molecule axes has been analysed. The axes are defined in Fig. 3.3, where a schematic view of two adjacent molecules from three different perspectives is shown with the blue circles marking the nitrogen positions. Two orthogonal molecule axes are defined: the light blue axis is called nitrogen axis, as it is defined by the nitrogen position of each imidazole molecule, and the light red axis is called methyl axis, as it is defined by the bond between the methyl group and the imidazole ring. The following arrangements between the two imidazole molecules are considered: both molecular axes parallel (0/0); 90° angle or 180° angle between the methyl axes with parallel nitrogen axes, noted (90/0) and (180/0); parallel methyl axes with 45° angle or 90° angle between the nitrogen axes, noted (0/45) and (0/90). In order to keep the orientation fixed during the minimisation, all nitrogen positions were fixed.

Results

The Fig. 3.4 shows the potential energy barrier for the proton transport. The barrier for imidazole obtained by DFTB is underestimated by about 2 kcal/mol compared to DFT(B3LYP functional and basis function set: cc-pVZQ for water and 6-31g* for sulphonic acid and imidazole). Proton transport in sulphonic acid systems and in water compares even better with the ab initio values. The energy barrier for all species increases with the distance between the groups, reaching 10 kcal/mol at about 0.30 nm for the imidazole system, at a shorter

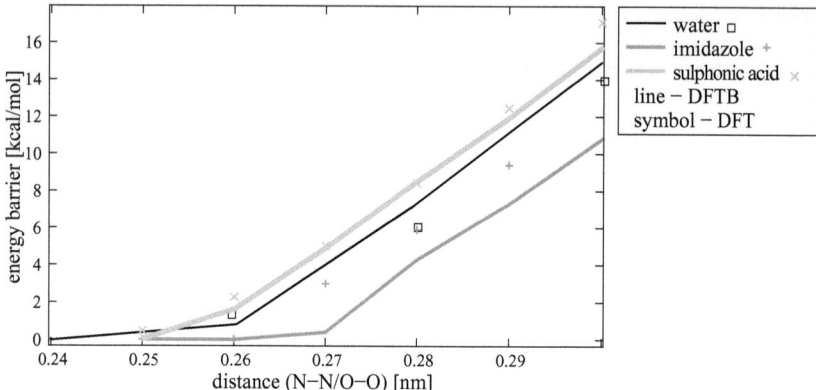

Figure 3.4: *Potential energy barrier for proton transport between two proton conducting species: water - black; methyl imidazole - dark grey; methyl sulphonic acid - light grey; straight lines derived from DFTB calculation, symbols from DFT calculation with B3LYP functional. The lines serve as guidance to the eye.*

distance than 0.29 nm for water and at about 0.28 nm for sulphonic acid. For each proton conducting species, a distance exists where the barrier vanishes: for imidazole at a nitrogen distance of about 0.27 nm, for sulphonic acid at an oxygen distance of about 0.25 nm, for water at an oxygen distance of about 0.26 nm. In water systems a positive charged complex of two water molecules bound via an additional interstitial hydrogen atom is called Zundel ion. Following this, the biatomic protonated configurations of imidazole and sulphonic acid are also called Zundel like configuration.

Zundel like imidazole complexes are already reported in the literature[76][73]. The intermolecular distance between nitrogen atoms is about 0.28 nm. The complex is chacterised by an elongation of the distance between the interstitial/intermediate hydrogen and the nitrogen atoms. The formation energy gain of a Zundel like imidazole complex is about 17 kcal/mol[i], while for water it is only 8 kcal/mol. For sulphonic acid two protonation states are considered, a dimer with a proton hole and a dimer with an excess proton. The gain is in both cases about 24 kcal/mol. The Zundel like aggregation of two protonated imidazole molecules involves the alignment of all nitrogen atoms and their bonded hydrogen atoms, as well as mirror symmetry at the plane orthogonal to the nitrogen axes through the interstitial hydrogen. The hydrogen is delocalised between two imidazole molecules. The angle between the imidazole molecules formed by the nitrogen atoms influences significantly the energy barrier for proton transport. The alignment of the axes favours proton transport, as reported in Fig. 3.4.

In Fig. 3.5, the minimal and maximal potential energy along the proton transport path is plotted. The abscissa is a measure for the fixed nitrogen distance and equals the distance between the position of the proton at the minimum and maximum of the energy barrier. Different arrangements of imidazole groups are compared. The alignment of imidazole means that both molecular axes are parallel, i.e. the nitrogen axis and the methyl axis as introduced

---

[i]A Zundel like imidazole complex is compared with an isolated neutral imidazole molecule and an isolated protonated imidazole molecule.

3.2. PROTON TRANSPORT ABILITY OF FUNCTIONAL GROUPS

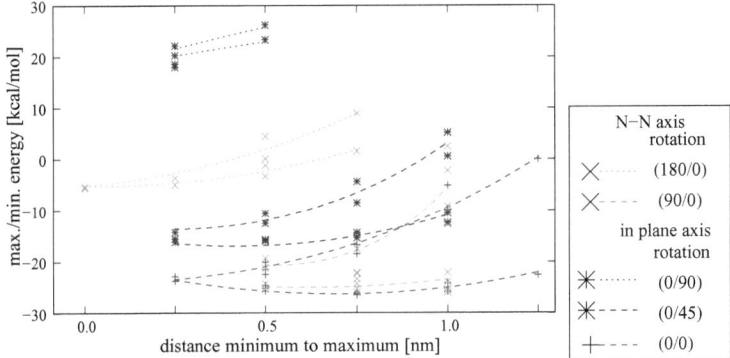

Figure 3.5: *The maximum and minimum potential energy for proton transport between two methyl imidazole molecules, depending on the alignment of the two rings. The angle between the nitrogen axes of both molecules and between the methyl axes are assigned in the figure, see Fig. . The lines serve as guidance to the eye.*

above. The rotation of the ring around the axes defined by the nitrogen atoms leads to an increase in total energy and transport barrier, as the formation of a Zundel like symmetric protonated complex is suppressed. Energy minimisation was performed under constraints of the nitrogen position. The aligned configuration and the configuration with an angle of 90° between the methyl axes show the lowest total energy. The slight increase of total energy for small distances in the aligned configuration is caused by the restraints, as the system is not allowed to relax freely and is hindered from forming the Zundel like configuration.

3.1.5 Short Conclusion

The underestimation of the binding energy per molecule for water clusters is up to 2 kcal/mol for H-bond-DFTB and 4 kcal/mol for primal DFTB with *pbc/mio* basis set. For the larger clusters, the H-bond-DFTB demonstrates its improvement of hydrogen bond description. The proton affinity is overestimated by up to 4 kcal/mol for the water clusters with both DFTB methods, while the proton affinity of imidazole and deprotonated sulphonic acid is underestimated by 6 to 7 kcal/mol. Proton affinity for larger clusters is described better than for smaller water clusters. The proton affinity of imidazole deviates by (-2)% for DFTB and (+1)% for DFT with cc-pVQZ basis from the proton affinity of a water cluster of size (n=10).

The deviation of potential energy barriers calculated with DFTB is less than 2 kcal/mol. For all systems, barrier free transport is observed for small distances corresponding to the formation of a Zundel like complex of the proton conducting species. For the imidazole case, the transport barrier and the formation of Zundel complex have been shown to depend on the alignment of the molecule due to its symmetry.

3.2 Proton Transport Ability of Functional Groups

In the following section the model systems of isolated proton conducting groups are introduced. These models serve to clarify the proton transport mechanism in a system of imidazole and sulphonic acid immobilised via carbon chain and to understand the influence of the geometrical arrangement of the functional molecules neglecting the influence of the substrate. In Sec 3.2.2 the diffusion coefficient in such imidazole systems is calculated and compared to the diffusion coefficients in liquid systems. In Sec. 3.2.3, structural properties and hydrogen bond networks are analysed and their influence on the proton transport is discussed for systems of isolated immobilised functional molecules.

3.2.1 Model System of Proton Conducting Groups in Vacuum

In the following, a very simple model system is considered consisting only of the functional groups, e.g. the proton conducting species - sulphonic acid or imidazole - covalently bound to a carbon chain, which is called spacer. The spacer has a length of 3 to 5 carbon atoms, according to Sec. 2.2. In most cases, pentyl sulphonic acid or pentyl imidazole is considered. While non-bonding interactions with the silicon dioxide material are neglected, its sustaining function is simulataed by using of restraints on the terminal ethyl group of the carbon chains, leaving an unrestrained propyl imidazole or propyl sulphonic acid group, see left part of Fig. 2.2. Besides pentyl, also heptyl and nonyl groups were considered that correspond to pentyl and heptyl spacer of the functionalised material in experiment. Restraining the terminal ethyl groups makes it possible to analyse the effect of immobilisation of the proton conducting groups due to the covalent bond and the density of groups on the proton transport.

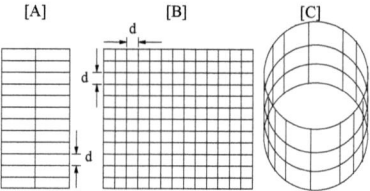

Figure 3.6: *Different models: groups ordered on a straight line (system A) or a flat surface (system B) with a defined distance d between the groups; groups ordered on a cylinder surface (system C).*

The system consists of pentyl sulphonic acid and pentyl imidazole in vacuum. Constraints on the terminal ethyl group of the carbon chains allows to equally space the molecules along a straight line (type A), on a flat regular two dimensional grid (type B) or on the inner surface of a cylinder barrel (type C) as assigned in Fig. 3.6. The system of type C geometrically corresponds to the functionalised porous structure while the other models are flat. In system C, the restrained ethyl groups are homogeneously distributed on a cylinder barrel with diameter of about 3 nm and a surface density of about 1 nm^{-2}. Note that the restraints on the system of type C lead to a diameter of 2.7 nm and an effective higher surface density of groups, as the effective radius of the barrel is reduced. Beside the chain length, the relevant properties of system A and system B are the surface density of functional groups or distance d between the restrained ethyl groups, therefore, in the following the notation (A/d) or (B/d) is used, where the variable d is the numerical value of the distance d as shown in Fig. 3.6.

3.2.2 Diffusion of an Excess Proton

The self diffusion coefficient of an excess proton is calculated for two different system types, model system A and B, consisting of propyl imidazole molecules with different distances between the groups. The distance (d) is in the range of 0.4 nm to 0.7 nm and the numbers of molecules in the simulations are 36 for system type B and 6, 12 or 18 for system type A, see Tab. 3.1.

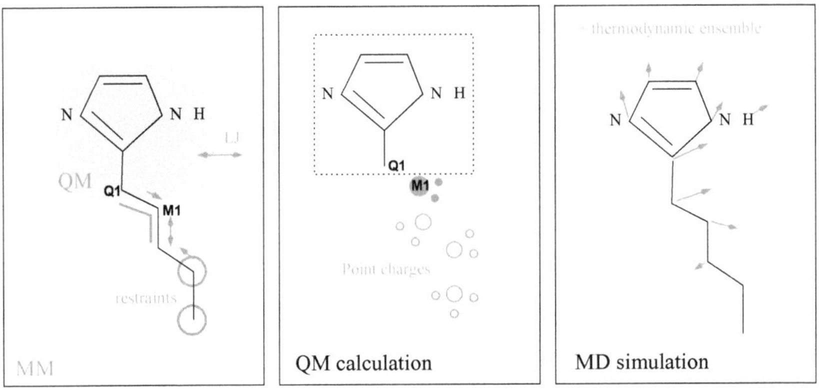

3.2. PROTON TRANSPORT ABILITY OF FUNCTIONAL GROUPS

Figure 3.7: *QM/MM coupling scheme for pentyl imidazole: The* **first** *picture shows the subdevision of the system in QM and MM part. Arrows symbolise the FF description as explained in Sec. 1.1.1. Restraints are considered. The* **second** *picture illustrates the QM calculations. MM atoms are included as an external field of point charges, thus electrostatic interactions between QM and MM parts are included. The* **third** *picture shows the evaluation of the MD simulation step. For further explanation, see Sec. 1.1.4.*

Computational Details

MD simulations were performed with a canonical ensemble (NVT) by use of the Berendsen thermostat with a time step of 1 fs at three different temperatures; $T = 350$ K, $T = 400$ K and $T = 450$ K. The QM/MM coupling schema was used, as described in Sec. 1.1.4. The QM part was described by primal DFTB with *pbc/mio* SK-file set and consisted of methyl imidazole groups with one hydrogen being the link atom, while the MM part was the remaining system, which was described by the OPLS-FF, see Fig. 3.7.

The schematic view in Fig. 3.7 illustrates the QM/MM coupling scheme. The first picture shows the subdevision of the system in QM and MM part, where arrows symbolise the FF description as explained in Sec. 1.1.1. According to Eq. 1.51, the MM calculation includes bonded and non bonded interactions between the MM atoms of the alkyl chains (H_{MM}^{total}), bonded interactions between the QM atom (Q1) and a MM atom (M1) at the boundary ($H_{QM/MM}^{bonded}$) and Lennard-Jones interaction (LJ) between QM and MM atoms (see $H_{QM/MM}^{LJ}$ in Eq. 1.51). The second picture shows the DFTB calculation of QM atoms in the statically electrical field which is created by the point charge distribution of the partial charges of all MM atoms. The partial charges of atoms close to a link atom are set to zero by distributing their charge to adjacent atoms to preserve charge neutrality.[i] In each time step of the MD simulation forces from the QM and the MM calculations on all atoms are evaluated and the dynamics is performed with respect to the thermodynamic ensemble, as shown in the third picture.

The diffusion coefficient (D) was calculated from the *mean squared displacement* (MSD) of the mCEC coordinate $\vec{c}(t)$ (see Sec.1.4) via the Einstein relation[118], which describes the behaviour of particles during Brownian motion in the limit of long time scales. In Eq. 3.5, the MSD was defined over the time interval $[t_0; t_0 + t_m]$, which was set to 50 ps. For a

---
[i]The partial charges of all atoms of the methyl group containing a M1 atom are set to zero.

number of molecules	parameter set distance d [nm]	Temperature [K]	diffusion D [10^{-8}m^{2}s^{-1}] 3-dim	±	experimental value [10^{-8}m^{2}s^{-1}]
\multicolumn{6}{c}{model system A}					
6	0.4	400	0.17		
12	0.4	400	0.13	0.01	
6	0.4	450	0.21		
12	0.5	400	0.08	0.01	
12	0.6	400	0.06	0.02	
6	0.7	400	0.08		
12	0.7	400	0.03	0.01	
18	0.7	400	0.04		
12	0.7	450	0.07	0.01	
\multicolumn{6}{c}{model system B}					
36	0.4	350	0.03		
36	0.4	400	0.12		
36	0.4	450	0.40		
\multicolumn{6}{c}{liquid proton conducting species}					
32	imidazole	400	0.13		≈ 0.2 [74]
32	imidazole	450	0.18		
	water	300	1.6		0.93 [66]

Table 3.1: *Self diffusion coefficients of an excess proton in different environment: model system A and B consisting on propyl imidazole molecules; liquid system: water and imidazole. The error in diffusion coefficient is only estimated for the systems containing 12 molecules.*

better statistical sampling, an ensemble average with respect to different trajectories was performed as well as an average with respect to the time origin t_0. According to the Einstein relation, the MSD shows a linear behaviour in the limit of long time t, with the diffusion coefficient being the slope in the linear part of the MSD. The linear behaviour of the MSD was considered after about 10 ps to 15 ps.

$$MSD = \frac{1}{6 N_{traj} (t_M - t_m)} \sum_{i=0}^{N_{traj}} \sum_{t_0=0}^{t_M - t_m} (\vec{c}_i(t + t_0) - \vec{c}_i(t_0))^2 , \qquad (3.5)$$

For the MSD, at least 5 trajectories (N_{traj}) with a production run time of 200 ps (t_M) were analysed for each parameter set of the model system A. The equilibration time was about 100 ps for each trajectory. The convergence of the analysis method was observed carefully. For model system B, 6 simulations of at least 50 ps were evaluated. As reference systems, liquid imidazole and liquid water were studied. Here, the entire system was described by DFTB method. The liquid imidazole (melted crystal model) consists of 32 molecules in a box resulting in a density of 11 nm^{-3} groups. 10 trajectories (N_{traj}) of 50 ps (t_M) (10 ps of equilibration before) for 2 different temperatures (400 K, 450 K) were analysed. For the water system, 10 trajectories of 50 ps were considered in a system size of 64 molecules.

3.2. PROTON TRANSPORT ABILITY OF FUNCTIONAL GROUPS

Self Diffusion Coefficient of the Excess Proton

In Table 3.1, the self diffusion coefficients of excess protons are listed for different model systems of propyl imidazole and liquid imidazole and water at different temperatures. Finite size effects occur for the small systems and for the bigger system, the error is estimated by the nitrogen diffusion for the systems containing 12 molecules.[i]. Taking into consideration the system A with d=0.7 nm, the system size in the super cell should be at least 12 molecules in a row[54]. Because of computational costs, system B was only analysed with a group distance of d=0.4 nm. But tendencies for systems of lower density are expected to be similar to those in systems of type A, as the formation of linear clusters is observed that may be described as well by the model system A, see Sec. 3.2.3. The calculation of diffusion constants needs very long trajectories, especially in the case of low constants as expected for larger distances.

The self diffusion coefficients of the excess proton in both liquids, imidazole and water, are in the same order of magnitude as the experimental values which are also listed in the table[74][66]. For water, the coefficient obtained by DFTB MD simulation is about a factor of 1.8 higher than the experimental value of $0.93 \cdot 10^{-8} m^2 s^{-1}$ at a temperature of 300 K. The deviation is probably caused by the underestimation of binding energies in DFTB, as discussed above for water clusters, which results in a faster formation and cleavage of hydrogen bonds and thus a faster proton transport.

For liquid imidazole, the simulated diffusion coefficients are $0.13 \cdot 10^{-8} m^2 s^{-1}$ (T=400 K) and $0.18 \cdot 10^{-8} m^2 s^{-1}$ (T=450 K). In liquids, proton diffusion has to two borderline mechanisms: the structural diffusion and the vehicular diffusion, see Sec. 2.1. In model A and B the proton conducting molecules are immobilised by the carbon chain, therefore, proton transport strongly depends on the hopping mechanism. The movement of protonated groups - thus the vehicular mechanism - is restricted by the flexibility of the carbon chain. The vehicular movement contributes only locally, depending on the distances between the groups and the length of the carbon chain, and contributions to the macroscopic diffusion are only indirectly, but at the same time essentially, bridging between two proton conducting molecules, especially at low density. System A and B with the highest surface density, d=0.4 nm, compare well with the system of liquid imidazole at a temperature of 400 K. Here, one may consider that the low distances in system A and B lead to a similar situation as in a liquid imidazole material, where vehicular mechanism has a minor influence. For a fluorinated imidazole system, vehicular diffusion is reported to contribute to 1/3 of the diffusion at $T = 450$ K, which is about $0.18 \cdot 10^{-9} m^2 s^{-1}$[67].

In all systems, the imidazole molecules are highly ordered due to hydrogen bond formation between the nitrogen and hydrogen atoms. During the MD run, Zundel like configurations - as discussed above - and a linear ordering of imidazole molecules in direction of the nitrogen axis are observed, compare Sec. 3.1.4. By increasing the distance between the groups in model system A from 0.4 nm to 0.7 nm, the diffusion constant decreases drastically by 75%, which can be interpreted as a reduction of the connected hydrogen bond network. At the same time, strong hydrogen bond interaction leads to immobilisation, which is only compensated for higher temperatures. Due to such temperature effect, the diffusion coefficient increases strongly comparing systems at 350 K, 400 K and 450 K, especially in the case of system type B, caused by increased dynamics of the system, i.e. by higher kinetic energy of the atoms.

---

[i]The nitrogen diffusion should be zero as the groups are immobilised.

3.2.3 Structural Properties

So far all systems ($d \leq 0.7$ nm) are at the high density limit of surface functionalisation as realised by synthesis, see Sec. 2.2. In the following, the parameter range is broadened, going to classical FF simulation. By using this method, charge transfer and reactions are no longer described and the analysis of the MD simulation does not lead to a value for the diffusion coefficient. So far, the trends in the diffusion coefficient for density and temperature dependence have been explained by changes in the hydrogen bond network. Since the vehicular transport mechanism is suppressed in the long range regime, in the following, a systematic analysis of MD simulations is performed in order to define the conditions that enable the proton hopping mechanism or the barrier free transport (see Sec. 3.1.4).

Computational Details

The structural properties of the different imidazole and sulphonic acid system are evaluated for two different temperatures; $T=300$ K and $T=450$ K. Therefore, classical MD simulations were performed in the canonical ensemble (NVT), using the Nose-Hoover thermostat with a coupling constant of about 0.1 ps and a time step of 1 fs. All atomic interactions are described by the OPLS FF and the model types are used as introduced in Sec. 3.2.1. The propyl imidazole systems are of type A and B with a distance between the functional molecules of $d=0.7$ nm and 1.0 nm, as well as a system of type C with a surface density of groups of $1.0/\text{nm}^2$. The propyl sulphonic acid systems are of type C with a density of $1.0/\text{nm}^2$ and of type B with a distance of $d=0.7$ nm, 1.0 nm, 1.4 nm and 2.0 nm. Furthermore, systems of type B with distance $d=1.0$ nm between different alkyl sulphonic acid groups are simulated. The number of the unrestrained carbon atoms of the alkane are 3 for the propyl, 5 for the heptyl and 7 for the nonyl group. Some systems are given in different protonation state, that means with one excess or defect proton in imidazole or sulphonic acid based systems, respectively. The number of molecules is 144 for system B, 24 for system A and 230 for system C. For comparison, the analysis of structural properties is not only performed for these systems, but also for the data from the QM/MM simulations, which have been explained in the previous paragraph.

Two different structural properties are calculated every thousandth step of the MD simulation which means every 1.0 ps the positions of the atoms are recorded and subsequently evaluated. For short distances, the average distance between the proton conducting species is calculated via RDF analysis, see Sec. 1.2.1. The number density, i.e. integral over the RDF, is used to analyse the structural properties of the system, such as the distance between proton conducting groups, between proton conducting groups and water or the surface. The RDF for nitrogen atoms (N) of imidazole or oxygen atoms (O) of the sulphonic acid were evaluated for trajectories of 1.0 ns after at last 1.0 ns equilibration. The RDF peak position and width were obtained from the fit of the peaks belonging to the next neighbour molecules by a Gaussian function $Ae^{-\frac{(x-x_0)^2}{2\sigma^2}}$. The choice of the functional form is arbitrary and facilitates the analysis. As the first peak is asymmetric, the two sides of the peak are fitted separately to obtain a different sigma for each side.

The long range ordering is inhomogeneous and dominated by the formation of aggregates of imidazole or sulphonic acid groups. An aggregate is defined as a set of proton conducting groups with a distance between at least two of the groups that is lower than the distance parameter c. The parameter c equals 0.35 nm in the case of imidazole and sulphonic acid represented by nitrogen and oxygen atoms, respectively. The size of the aggregate is given by the number of component groups. The probability that an arbitrary group is a mem-

3.2. PROTON TRANSPORT ABILITY OF FUNCTIONAL GROUPS

ber of an aggregate of size N is called the distribution of the aggregate of size N. Another important property - beside the size - is the stability of the aggregate. The analysis of the stability is performed by analysing the ratio (percentage) of clusters that exist longer than the time $\Delta t=25$ ps. During this time, the aggregate is not changing - neither in size nor in the elements.

Local Distribution of Proton Conducting Species

Table 3.2: *First part: N-N RDF form FF-MD at 400 K for different imidazole model systems: A, B, C. Second part: data marked by (*) from QM/MM - MD as comparison, imidazole system A. Third part: O-O RDF form FF-MD at 400 K for different sulphonic acid model systems: A, B, C. The protonation state (+/-1) signifies that the RDF contains one protonated imidazole group or one deprotonated sulphonic acid group. The carbon chain length is given in number of unrestrained carbon atoms: 3 -pentyl; 5 -heptyl; 7 -nonyl group. Both peaks are fitted to a gaussian function $Ae^{\frac{(x-x_0)^2}{2\sigma^2}}$.*

model	d [nm]	carbon chain	protonation state	Peak 1 x_0 [nm]	$\sigma_1^{(-)}$	$\sigma_1^{(+)}$	Peak 2 x_0 [nm]	σ_2
				imidazole system				
A	1.0	3		0.297	-0.013	0.032	0.505	0.040
A	1.0	3	+1	0.277	-0.009	0.018	0.486	0.020
A	0.7	3		0.301	-0.017	0.026	0.511	0.038
A	0.7	3	+1	0.283	-0.013	0.018	0.492	0.020
B	1.0	3		0.290	-0.014	0.028	0.497	0.032
B	1.0	3	+1	0.284	-0.013	0.017	0.490	0.026
B	0.7	3		0.295	-0.013	0.021	0.500	0.029
B	0.7	3	+1	0.283	-0.010	0.016	0.490	0.027
C				0.293	-0.012	0.017	0.505	0.018
			imidazole system (QM/MM, see Sec. 3.2.2)					
A*	0.7	3	0/+1	ca.0.28(0.26)			ca.0.49	
B*	0.7	3	0/+1	ca.0.29(0.26)			ca.0.50	
			sulphonic acid system					
B	0.7	3		0.279	-0.013	0.018		
B	0.7	3	-1	0.271	-0.010	0.014		
B	1.0	3	-1	0.271	-0.011	0.015		
B	1.4	3		0.286	-0.015	0.042		
B	1.4	3	-1	0.270	-0.010	0.014		
B	2.0	3	0 /-1	1.998	-0.463	0.675		
B	1.0	5		0.288	-0.013	0.024		
B	1.0	5	-1	0.288	-0.013	0.019		
B	1.0	7		0.278	-0.014	0.025		
B	1.0	7	-1	0.276	-0.012	0.020		

Previously, a structural transport was observed involving a crossover of hydrogen bonds to covalent bonds and vice versa between nitrogen and hydrogen atoms, which was identified with the barrier free proton as described in Sec. 3.1.4. This hopping event demands high local densities of proton conducting groups in systems of relatively low average densities compared to liquids. The RDF between proton conducting species is a measure for their local density corresponding to the average distance. In Tab. 3.2, peak positions of the RDF between nitrogen atoms of imidazole molecules or sulphur atoms of sulphonic acid molecules are reported for different systems of functional molecules in vacuum. The functional molecules are homogeneously arranged on a straight line (system A), on a flat surface (system B) or on a cylinder surface (system C), as assigned in Fig. 3.6. The peaks are charcterised by their position x_0 and their width σ, which are obtained from fitting a Gaussian function to the peaks. As the first N-N peak is strongly asymmetric, two widths are listed in the table.

For all imidazole systems, the first peak is in the range of 0.277 nm to 0.301 nm, while the second peak is shifted by about 0.21 nm, which equals the intramolecular distance between nitrogen atoms. Therefore, a sharp second peak indicates an alignment of the imidazole groups[54]. For sulphonic acid the peak is located at 0.270 nm to 0.288 nm, for all systems except for the system B with d=2.0 nm. For (B/2.0 nm), non-bonded interactions are negligible due to the length of the flexible part of the alkyl chain, about (0.7±0.1) nm. As a consequence, the first peak is located at about the periodicity and a broadening of the peak is observed.

For comparison, also QM/MM propyl imidazole systems (as described in Sec. 3.2.2) are included in the analysis, marked by a star in Tab. 3.2. For the QM/MM systems, a split of the first peak is observed. The higher maximum marks the next neighbour distances of neutral molecules at about 0.28 to 0.29 nm, while the smaller maximum is related to the formation of a Zundel like complex with a reduced N-N distance of approximately 0.27 nm, as discussed above. The FF simulation shows a similar difference for protonated and unprotonated systems. In the case of pentyl imidazole systems with next neighbour distances of 0.7 to 1.0 nm, the position of the first peak varies between 0.277 and 0.284 nm for RDF between protonated groups and their neighbours, while it is located between 0.290 and 0.301 nm for neutral imidazole systems. The widths of the first and second peak become smaller for the protonated group comparing to the same model systems. Similar trends are observed for deprotonated pentyl sulphonic acid, here the position of the first peak is shifted from 0.28 nm to 0.27 nm for the deprotonated molecule. Even for systems with a next neighbour distance up to 1.4 nm, the same shift is observed and all neutral systems show a broader peak. In the case of sulphonic acid systems, the chain length was increased, which leads to a broadening and shift to higher distances of the first peak. This behaviour is explained by the formation of bigger aggregates as discussed in Sec. 3.2.3.

The topological effect on the next neighbour molecules is analysed for the imidazole system. As the peak position is the same for system C, both systems B and both systems A, the topological effect is negligible on the small distance scale and only differences in the peak width indicate differences in the distribution of proton conducting groups. Therefore, the integral over the RDF is calculated in order to obtain the number density of the species over the distance. In Fig. 3.8, the integral is plotted for imidazole. As discussed for the RDF, stepwise increase of the curve is observed at about 0.29 nm and 0.5 nm. For system (A/0.7 nm), (B/0.7 nm) and (B/1.0 nm), the protonated system shows a number density of 2, which means that on average one nitrogen atom (of the neighbouring molecule) is found in about 0.3 nm close to each nitrogen of the protonated molecule. In the case of system (A/1.0 nm) the number density is significantly lower. The dotted line represents the number density of neutral molecules, which shows significantly lower values, consistent with the shift in the RDF peak. For system C, only the neutral case was analysed.

A similar analysis is performed for sulphonic acid systems, plotted in Fig. 3.9. All propyl sulphonic acid systems show an increase to the number density of 3 at a distance of 0.5 nm, which means that 3 adjacent sulphur atoms are averagely in a distance of about 0.5 nm. For systems with a longer alkyl chain, the first step is lower, which follows directly from the

3.2. PROTON TRANSPORT ABILITY OF FUNCTIONAL GROUPS

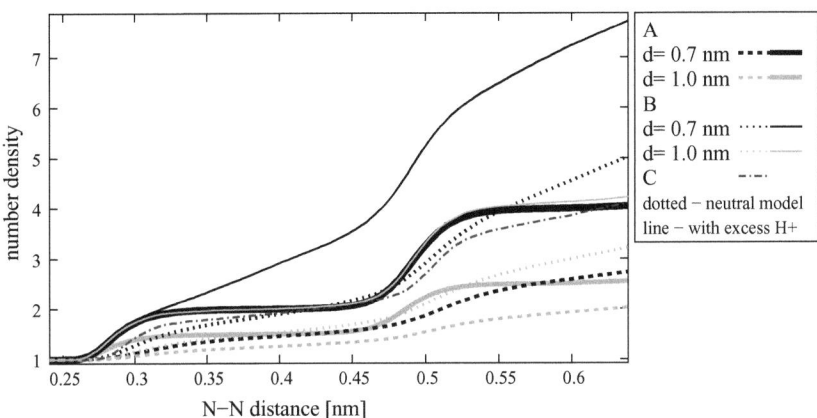

Figure 3.8: *Number density of imidazole for different model system (A - thick line, B - thin line and C): distance between next neighbour groups (d = 0.7 nm - black; d = 1.00 nm - grey), model contains excess proton - straight line, without - dotted.*

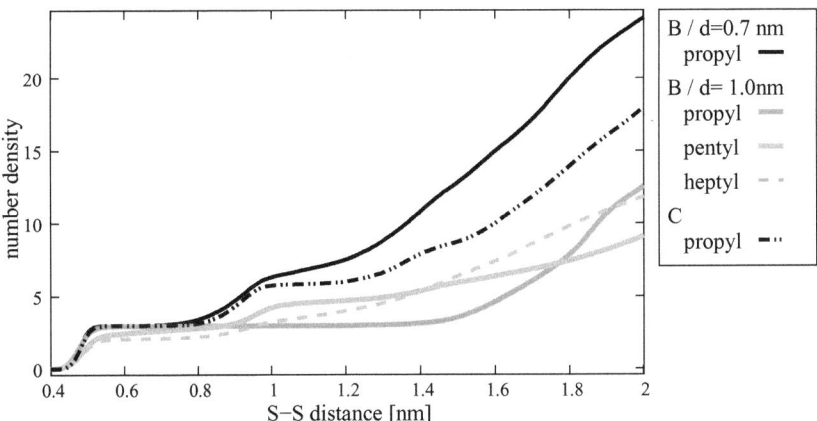

Figure 3.9: *Number density of sulphonic atoms for different system of propyl sulphonic acid molecules; C and B: distance between next neighbour groups (d = 0.7 nm - black; d = 1.00 nm - grey) and system of pentyl sulphonic acid molecules for d = 1.00 - dashed.*

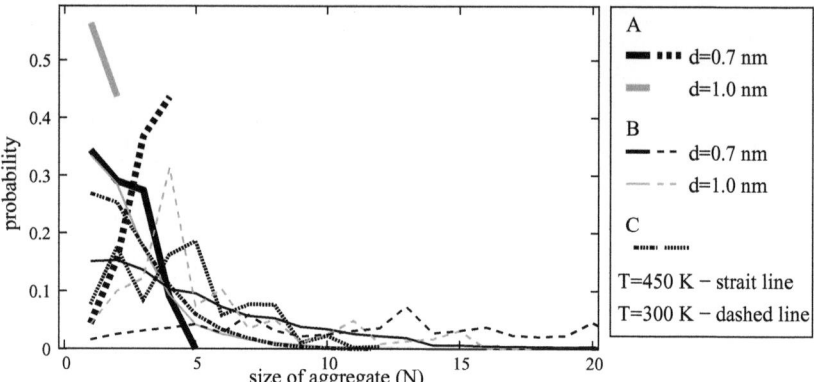

Figure 3.10: *Probability to find an arbitrary imidazole group as part of the aggregate with size N. Comparison between different model systems and temperatures: thin and thick black line - systems types B and A with distance 0.7 nm; thin and thick grey line - system type B and A with distance 1.0 nm; black dashed line - system type C. Temperatures: T =300 K - straight line; T =450 K - dashed line.*

broadening of the first peak that was observed for the RDF before. The longer the chain, the more suppressed are the steps in the number density, which hints to the formation of big aggregates as well as a sterical hindrance by the alkyl group. For the systems (B/0.7 nm), (C) and (B/1.0 nm with pentyl chain), a second step at 1.0 nm is observed, which marks the distance to the second next neighbour molecules.

For both functional groups, propyl imidazole and propyl sulphonic acid, the system C shows a number density similar to the (B/0.7 nm) system up to a distance of 1 nm. For properties at these distances, the use of a flat functionalised surface is justified. The surface density of functional groups in model system C is about 1 nm^{-2} with respect to the restrained carbon atoms, while the surface density inside the cylinder is already higher for the first unrestrained carbon atoms due to an effectively reduced pore radius. This obviously results in a higher density of the corresponding flat system.

Particle Aggregation - Global Ordering

As the long range order of the proton conducting groups is inhomogeneously structured, the aggregation of imidazole or sulphonic acid groups is analysed in the MD simulations. Large aggregates correspond to extensive regions with low proton transport barriers, as the proton conducting groups in these regions are sufficiently close to each other.

The distribution of aggregates, as shown in Fig. 3.10 and 3.11, describes the probability that an arbitrary group takes part in an aggregate with size N. For the imidazole functionalised sytem, the influence of the system type is studied. Systems of type B have larger aggregates than systems of type A, as the distribution is broadened significantly. Furthermore, higher density of groups leads to broader distribution and lower temperature increases the aggregate size. Moreover, longer chain length lead to larger clusters as shown for the sulphonic acid systems. An increase in density leads to a decrease in the maximum of the cluster distribution, as well as in the probability to find a solitary group due to a broader

3.2. PROTON TRANSPORT ABILITY OF FUNCTIONAL GROUPS

distribution.

In the RDF analysis, the type C model compared well with (B/0.7 nm) at short distances for both proton conducting species. For the sulphonic acid system, the approximation is supported also by the characteristic of the aggregate distribution, see Fig. 3.11. Fig. 3.10 shows instead a better aggreement between the aggregate distributions of system C and system (B/1.0 nm) for imidazole. One notices that the largest aggregates in system C are of size 10, while the size in system (B/0.7 nm) is not limited to such small numbers. The deviation in the distribution probably results from the alignment of adjacent imidazole molecules. The curvature of the surface prevents the formation of linearly oriented aggregates of such size, which highlights the limitation of the flat system on a nanometer scale. As mentioned above, at small distances up to 1.0 nm the approximation of the cylindrical, functionalised environment by a flat model is still justified.

In Tab. 3.3, the maximum in the distribution and its height are listed for each system.

Table 3.3: *Aggregates of the most probable size (N) and the probability distribution are given as well as the probability to find those aggregates for more than 25.0 ps.*

model	d [nm]	temperature [K]	max. in aggregate N	[%]	stability 25 ps [%]	isolated group [%]	stability 25 ps [%]
imidazole							
A	0.7	300	4	53	5	8	8
A	1.0	300	2	84	64	16	61
B	0.7	300	7	6	0	2	25
B	1.0	300	4	24	17	4	19
C		300	4	19		8	10
A	0.7	450	1	40	9	40	9
A	1.0	450	1	62	43	62	43
B	0.7	450	1	13	5	13	5
B	1.0	450	1	36	18	36	18
B	1.4	450	1	100	100	100	100
C		450	1	29	10	29	10
sulphonic acid							
B	1.0	300	4	58	99	1	100
B	1.0	450	4	60	96	1	32
B	1.4	450	2	82	91	18	87
B(5+2)	1.0	450	4	25	81	2	0
B(7+2)	1.0	450	4	13	82	1	20

The probability to find an isolated group is also given. Comparing the pentyl imidazole and the pentyl sulphonic acid system, one notices that the sulphonic acid systems have larger aggregates, especially for system (B/1.0 nm). In general, the maximum in the distribution is at the isolated group for the imidazole system at the higher temperature, while for sulphonic acid the maximum is at an aggregate size of 4 groups in most cases, even at both temperatures. In the case of (B/0.7 nm) for sulphonic acid the maximum is not given, since the distribution is very broad.

The differences between the functional groups arise from the fact that the number of hydrogen bonds per group is higher in the sulphonic acid case than in the imidazole case. The imidazole groups show a linear arrangement, as hydrogen bonds only form between the two

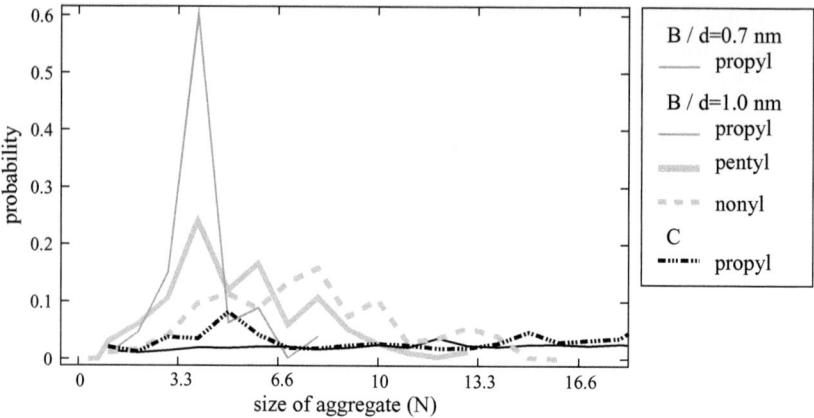

Figure 3.11: *Probability to find an arbitrary sulphonic acid group as part of the aggregate with size N at T =450 K; in system of type (B/1.0 nm) - grey: different chain length, heptyl (with (7+2)C flexible/ unrestrained), pentyl (with (5+2)C flexible/ unrestrained) and propyl group (with (3+2)C flexible/ unrestrained); system of type (B/0.7 nm) - black; system of type C - black dashed line.*

nitrogen atoms of each molecule (N-H and N), while sulphonic acid has 3 oxygen atoms to build hydrogen bonds with neighbouring groups. Therefore, the clusters are spherical and more stable. Note that the linear aggregate of imidazole bridges larger distances than the spherical cluster with the same number of sulphonic acid groups.

Beside the formation of aggregates, the hydrogen bond fluctuation is an important property influencing proton transport. An adequate measure is the stability of aggregates, which is also given in Tab. 3.3. The fractions of stable aggregates over 25 ps are listed for both aggregates, the maximum in the distribution and the aggregate with size N=1, which refers to an isolated group. For the isolated groups, stability means that these groups are isolated for more than 25 ps. If a group does not approach another group on this time scale, it is considered not to influence the proton transport significantly and therefore, it is called an inactive group on the time scale of 25 ps. The percentage of inactive groups is given by the percentage of isolated groups times the fraction of stable groups. The time scale is chosen in comparison with the relevant time scales in the imidazole crystal at T=390 K [73]. Proton hopping takes place in about 0.3 ps and reorientation of the molecules in about 30 ps. Taking this picture into account, fluctuations on the time scale of picoseconds enable the restrained vehicular mechanism by the carbon chain dynamic and the formation of new aggregates.

For imidazole, the isolated groups are considered. Systems of type A show averagely more fluctuations than systems of type B with the same distance d. A higher density of groups leads to lower stability of the isolated groups. Looking at the sulphonic acid system, the stability of the aggregates of size 4 is compared. They are in general much more stable than the isolated imidazole group. For all listed system at 300 K, the percentage of inactive groups is less than 1% except for system (A/1.0 nm) which results in 10% inactive groups for imidazole. At 450 K in imidazole system, there are 6% of inactive groups in system (B/1.0 nm), 26% in system (A/1.0 nm) and 3% in system (A/0.7 nm). For all other systems with $d = 0.7$ nm or $d = 1.0$ nm, the number of inactive groups are below 1%. Especially for the sulphonic acid

3.2. PROTON TRANSPORT ABILITY OF FUNCTIONAL GROUPS

systems, there are no inactive groups.

The distances between the aggregates are at least about the average distances between the groups. The proton transport on elongated distances takes place by rearrangement of the aggregates. This means that aggregate instability enables the transport from one cluster to the adjacent aggregate on the considered time scale. The instability is adequate to the fluctuation of hydrogen bonds.

3.2.4 Short Conclusion

The self diffusion coefficients, obtained by DFTB MD simulations, of an excess proton in liquid imidazole and in water are in the same order of magnitude as diffusion coefficients from measurements. In a system with a high surface density of functional imidazole molecules in vacuum, the proton diffusion coefficient is in the range of $10^{-9} m^2 s^{-1}$, which is similar to diffusion in a system of liquid imidazole. In liquids two borderline proton transport mechanisms exist, the structural transport and the vehicular transport, which can be considered to some extent independently in these systems; while the structural as well as the vehicular mechanism depend on each other to enable long range proton transport in a system of proton conducting species anchored by a spacer.

The FF simulations are used to scan a broader parameter range. The formation of hydrogen bonds between the proton conducting groups is observed. In parallel with the diffusion coefficient analysis, two conflicting factors are pointed out that favour proton transport: the formation of large aggregates and the fluctuation of hydrogen bonds. The first factor creates regions with low proton barriers, which favours the hopping mechanism, while the second enables the chain movement, i.e. the local vehicular transport. The diffusion coefficient is increased by higher density of groups, which corresponds to a higher probability of forming aggregates in the FF model. Furthermore, a higher temperature has an enhancing effect on diffusion, even though the aggregates analysis shows an increase of isolated groups. The temperature effect is explained by the lower stability of aggregates, especially isolated groups, due to hydrogen bond fluctuations. For imidazole at 450 K the (local)vehicular like transport is more probable, while for sulphonic acid at both temperatures quite stable aggregates are formed and the number of isolated groups is small.

The influence of the geometrical arrangement was analysed comparing the different FF models A, B and C. RDF analysis shows no qualitative difference between the cylindrical and the planar arrangement of functional molecules and a mapping of C to (B/0.7) is suggested. The surface density of system B is higher, as the curvature of the surface results in an effective higher surface density in systems of type C with density 1.0 nm^{-2}. Therefore, in the following the environmental influence on the proton transport mechanism of the anchored groups is analysed for a slab model system.

Chapter 4

Functionalised Silicon Dioxide Material

Being part of an interdisciplinary reasearch project, the PEM additives in focus comply with findings from experiment and from computational simulations as well as with the boundaries of chemical synthesis, as described in Sec. 2.2. In Sec. 4.1.1, a model system of amorphous silicon dioxide is introduced and the different parameters of the material are mentioned. Using the adequate model, two different types of classical MD simulations are performed. The first one aims to determine the maximal water content at ambient conditions, see Sec. 4.2, and the second one is performed at different humidity conditions, see Sec. 4.3. The chemical environment is characterised for different humidity by the following properties: the density profile inside the pore, the water environment and the interaction of groups with the surface.

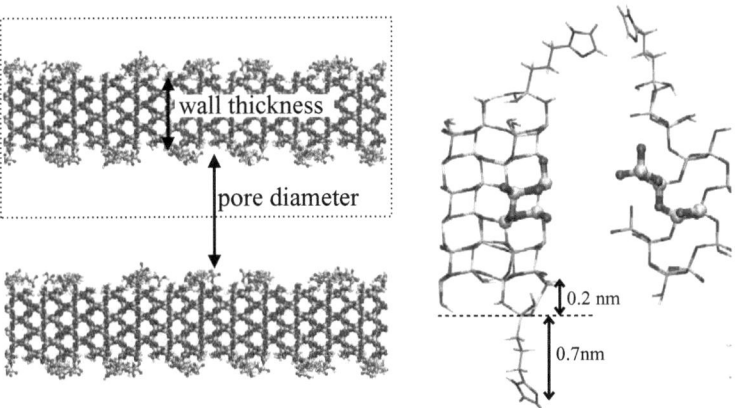

Figure 4.1: *Imidazole functionalised cristobalite slab model, periodic box marked by dotted line; close vision of imidazole functionalised cristobalite, atoms part of repeating unit marked by balls; dark grey - oxygen, light grey - silicon, white - hydrogen; further nitrogen and carbon atoms in the functional groups.*

4.1 The Amorphous Silica Model and FF Simulations

The PEM additives consist of functionalised MCM-41, as described in Sec. 2.2.1. The amorphous silicon dioxide material of 2.2 g cm^{-3} density is arranged in a porous structure with about 3.0 nm pore diameter. The suface density of hydroxyl groups in MCM-41 is estimated by different experimental and computational methods as ranging from 2 to 6 groups nm^{-2}. The density of functional groups obtained by experiment is so far limited to about 1.0 group nm^{-2}. In the following, an adequate model of the amourphous silicon dioxide material is introduced and computational details of the simulations are given.

4.1.1 Model System

The tetragonal crystal α cristobalite is a high temperature modification of silicon dioxide with a density of 2.2 g/cm^3 which is similar to the density of the amorphous silicon dioxide in the MCM-41. As long range periodicity has a minor influence on the surface interaction[119], a slab model based on the cristobalite crystal is expected to adequately represent most properties of the amorphous silica surface[120].

To build up the slab model of the crystalline material, the periodicity was broken in the (100) direction and the cristobalite surface was fully saturated by hydroxyl groups. The resulting surface density of about 6 nm^{-2} groups is in the upper range of estimates for MCM-41. The diameter of the slab was chosen according to the experimental wall thickness of the pores in MCM-41, i.e. about 1.0 nm. By changing the super cell size perpendicular to the surface, flat pore models with differing wall distances were obtained easily. According to Sec. 2.2.1, experimental values for the pore diameter are at the nanometres size, see Fig. 4.1.

So far, the unfunctionalised slab model has been described, which is only characterised by different pore diameters. All other parameters are unchanged here. Further properties, such as the surface densities of two different functional groups, are introduced corresponding to the functionalisations described in Chapter 3. The organic functional molecules consist of dihydroxyl silane propyl imidazole or dihydroxyl silane propyl sulphonic acid, which are covalently bound to the silicon dioxide surface. In accordance with surface functionalisation via a condensation reaction, the functional molecules replace a surface hydroxyl group in the slab model.

In the case of the sulphonic acid functionalised slab model, two states of protonation have to be considered. As discussed also later, deprotonation of the acid is expected under hydration, thereby introducing intrinsic charge carriers, e.g. hydronium ions by FF description, into the system. In the FF simulation, both types of sulphonic acid are considered: One system with completely deprotonated, negatively charged SO_3^- groups and hydronium ions in the water to ensure neutrality; another system with only neutral sulphonic acid groups. Both systems represent limiting cases of fully hydrated and completely dry systems. Since a dynamic equilibrium between protonated and deprotonated sulphonic acid is not possible by the FF description, both limiting system types were chosen for several hydration levels, while intermediate system types are neglected.

4.1.2 Computational Details

Classical MD simulations are performed, as described in Sec. 1.2. The model systems were built from the cristobalite slab model as introduced previously. During the simulations, rotations of the slab model were prevented by positional restraints on three of the silicon atoms in a fully periodic system. All interactions in the system were described by classical

4.1. THE AMORPHOUS SILICA MODEL AND FF SIMULATIONS

FF parameters taken from the literature. As a FF water model TIP3P[121] was applied. To describe the silicon dioxide, the FF of Lopes et al. was employed, as it has been parameterised for the interaction with organic molecules and TIP3P water[122]. In consistency with Chapter 3, the interaction of organic molecules is described by the OPLS FF (Sec. 3.2.2). All FF parameters are listed in the Appendix.

Simulation (I)

In the first step, the water contents of pores at full hydration are determined. Therefore, classical MD simulations were carried out in the NPT ensemble. The Nosé-Hoover thermostat with a time constant of 0.1 ps was applied for reference temperatures of 300 K, 400 K or 450 K. The slab model of the pore - as described above - was modified in order to bring it into contact with a water reservoir. In one direction parallel to the slab surface, the slab is infinitely continued by its periodic image, while the periodicity was broken in the other direction and all broken bonds were saturated with hydroxyl groups. The length of the water reservoir equals about three times the length of the slab, which leads to a total length of about 30 nm, see Fig. 4.2. To obtain equilibrium conditions between the water reservoir and the silicon

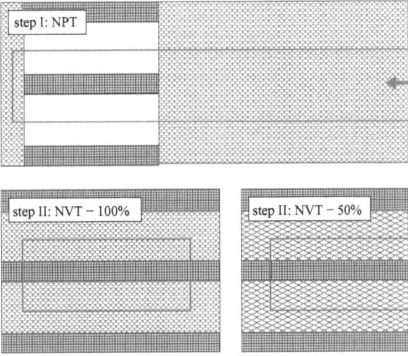

Figure 4.2: MD simulation, first step in NVP ensemble, second step in NVT ensemble; arrow - changes in length of boundary box vector.

dioxide region in the NPT ensemble, the anisotropic Berendsen barostat with a time step of 0.01 ps was applied to ensure a pressure of 1 bar while keeping the box vector orthogonal to the surface unchanged, as this direction refers to the pore diameter.

Four types of surface coverage were considered for this kind of simulation: unfunctionalised surface covered only by hydroxyl groups, (deprotonated) sulphonic acid functionalised surface covered with 1.3 groups nm^{-2} and finally imidazole functionalised surface covered with 1.3 or 0.65 groups nm^{-2}. The simulation aims at determining the limit of a fully hydrated system. Therefore, all sulphonic acid groups are in the deprotonated state. The pore diameter was set to four different values for the functionalised slab models with 1.3 nm^{-2} density, 1.5 nm, 2.0 nm, 2.5 nm and 3.0 nm. For all other systems the diameter corresponded to the pore diameter in MCM-41, i.e. 3 nm.

Simulation (II)

In a second step, different humidity conditions were simulated on the slab models as described in Sec. 4.1. The slab models were attached to themselves via the periodic boundary conditions in both dimensions parallel to the surface and different numbers of water molecules were introduced into the system, see Fig. 4.2. The contents inside the functionalised pores range between 100 % and 10 % as determined by simulation (I) for the corresponding models. Four different systems were considered: the unfunctionalised slab model, the imidazole functionalised slab model, the sulphonic acid functionalised slab model and the sulphonic acid functionalised slab model in the deprotonated state. The density of functional groups was set to 1.3 nm^{-2} and the pore diameter was set to 3 nm for all models.

MD simulations in the canonical ensemble were performed at 300 K and 450 K using a Nosé-Hoover thermostat with a time constant of 0.1 ps. The local environment inside the pores is evaluated after at least 1 ns of equilibration. The length of the production run is about 1 ns. For the analysis of the MD output, the trajectories were collected at every picosecond. Part of the analysis of the data is presented in the next chapter according to the thematical ordering.

4.2 Fully Hydrated Environment

In order to get an insight in the chemical environment of the functionalised silicon dioxide particles, the maximal water content is determined at ambient conditions by simulation of type (I) as described above. Fed from the reservoir, the water uptake of the slab model was evaluated after equilibration by two different parameters, the density and the commonly used parameter λ describing the number of water molecules per functional group.

4.2.1 Average Water Density

To avoid effects of the interface with the water reservoir of simulation (I), the water densities inside the pore were evaluated in volumes defined by both surfaces and two planes parallel to the interface at a distance of about 1.0 nm from the water reservoir. The first silicon layer acted as a reference for the surface of the interface slab models. The water densities were time averaged over 1.0 ns production runs in equilibrated systems.

Table 4.1: *Water content inside the functionalised slab models at different temperatures, surface densitis of propyl-imidazole or propyl-sulphonic acid and pore diameters, i.e. distances of slab surface.*

distance nm	density of groups [nm^{-2}]	water, density [g cm^{-3}] 300 K	400 K	450 K	correction volume [%]	density [%]
3.0	0	1.0	0.87	0.81	0	0
Imidazole Functionalised System							
3.0	0.65	0.89	0.78	0.73	-7	7
3.0	1.3	0.91	0.82	0.73	-14	16
2.5	1.3	0.89	-	0.69	-17	20
2.0	1.3	0.81	-	0.63	-21	26
1.5	1.3	0.66	-	0.54	-28	38
Sulphonic Acid Functionalised System							
3.0	0.65	0.95	0.9	0.80	-6	6
2.5	0.65	0.96	0.93	0.78	-8	9
2.0	0.65	0.93	0.9	0.78	-9	10
1.5	0.65	0.84	0.81	0.75	-12	14

The water densities inside different slab model pores are listed in Tab. 4.1. The porous environments were in equilibrium with a water reservoir at three different temperatures. In the case of an unfunctionalised - but fully hydroxylated - pore of 3.0 nm diameter, the average water density of 1.0 g cm^{-3} at 300 K is comparable to the density of bulk water. Going to functionalised pore models, the water uptake is decreased, since the actual pore volume is decreased by the volume occupied by the functional molecules, which was estimated by the

4.2. FULLY HYDRATED ENVIRONMENT

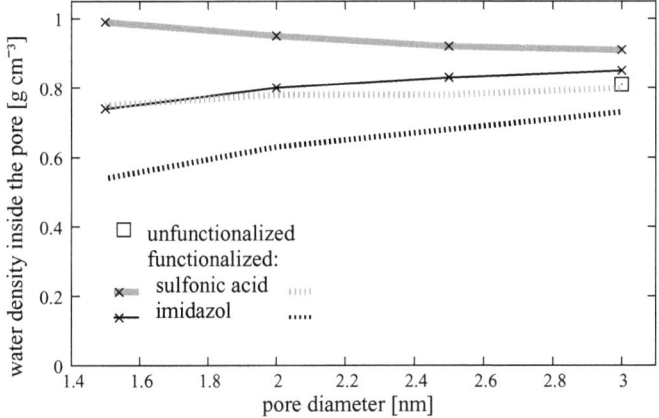

Figure 4.3: *Water uptake of slab model, dotted line - density and straight line - corrected density, spheres of van der Waals radii are subtracted; MD simulation at T=450 K for system and the unfunctionalised model.*

Van der Waals radii of the atoms of the dihydroxy silane propyl imidazole or dihydroxy silane propyl sulphonic acid. The overlap of the spheres - due to chemical bond between two atoms - was subtracted from the total volume to obtain the volume of the entire molecule. The influence of bond angles was neglected in these estimates. In the case of dihydroxy silane propyl imidazole, a volume of 0.16 nm^3 per molecule was calculated[i]. In the case of dihydroxy silane propyl sulphonic acid, a volume of 0.14 nm^3 was considered[ii]. The correction of the volume partially cancels the systematic error. The correction is also given in the table and results in a change of density by 10 % to 40 %.

The corrected and non-corrected densities should be considered together for a careful interpretation, as also the volume estimate has a high uncertainty. For example the volume of hydroxyl groups were neglected. With approximately 0.02 nm^3 per OH-group, the water density of the unfunctionalised slab is changed by 4 % to 9 %, as it lowers the volume by 4 % to 8 % for pores of 3 nm and 1.5 nm, respectively.

Figure 4.3 shows the densities and the corrected densities at a temperature of 450 K for different pore diameters, as listed in Tab. 4.1. The correction of the density slightly overcancels the deviation between functionalised and unfunctionalised pores, as it leads to higher density than in the unfunctionalised case. The sulphonic acid functionalisation shows the highest corrected water density, while the imidazole deviates only slightly from the unfunctionalised model. For smaller pores the differences for both models get even more pronounced. The sulphonic acid groups are obviously more hydrophilic than the imidazole groups. Comparing lower and higher surface functionalisations from Tab. 4.1, more imidazole groups lead

---
[i]Since the molecule is composed of 11 spheres with radius r_H=0.12 nm, 6 spheres with radius r_C=0.17 nm, 2 spheres with radius r_N=0.155 nm, 2 spheres with radius r_O=0.152 nm and one sphere with radius r_{Si}=0.21 nm.

[ii]The occupied volume is formed by 8 spheres with radius r_H=0.12 nm, 3 spheres with radius r_C=0.17 nm, one sphere with radius r_S=0.180 nm, 5 spheres with radius r_O=0.152 nm and one sphere with radius r_{Si}=0.21 nm.

to higher corrected densities at all temperature. As this trend opposes the trend observed for minimisation of the pore size this gives an estimate for the error range of water density of about 5 %.

As expected, the water uptake decreases by about 10 % and 20 % for the imidazole model when increasing the temperature to 400 K and 450 K, respectively, for the 3.0 nm pore. The loss in the unfunctionalised system and the system with only 0.65 imidazole groups nm^{-2} is similar (13 % to 12 % and 19 % to 18 %), whereas the decrease for sulphonic acid functionalised models is significantly lower, about 5 % and 16 % in the same temperature range for 3.0 nm pores. This means that the more hydrophilic system retains more water at higher temperature. The latter observation allows us to consider changes in the water distribution due to surface density of imidazole as insignificant. In contrast to this, the functionalisation with sulphonic acid groups enhances the water interaction with the inorganic material. This argumentation is independent from the volume effect of the different systems, as systems of the same type are compared.

For the high temperature case, one has to take into account that the pore model is still coupled to the water reservoir and that barostating to a pressure of 1 bar was applied. Therefore, evaporation of the water is not possible, whereas this is a major problem under fuel cell conditions as mentioned before. The total pressure in the system equals the partial pressure of water as no other species were included in the simulations. Therefore, the calculated water densities are considered as a fully hydrated case comparable to the high partial pressure regime of an experimental water adsorption isotherm.

4.2.2 Average Number of Water Molecules per Group

Beside the water density, the number of water molecules per functional group is determined. This parameter is often symbolised by λ and is obtained by dividing the number of water molecules in a section of the pore wall surface. In the imidazole case, full hydration is referred to for a pore filling equivalent to 29 and 36 number of water molecules per functional group (λ) at 450 K and 300 K, respectively. In the sulphonic acid case, λ is slightly higher with 31 and 37 water molecules per group at both temperatures.

4.3 Different Humidity

In the next step, different hydration conditions are considered and the local environment in these conditions is estimated from simulations of type (II). The 1.3 nm^{-2} functionalised 3 nm pore models are filled with different numbers of water molecules, corresponding to percentages of λ for the fully hydrated case as determined by the previous simulations.

4.3.1 Density Profile

The density profiles perpendicular to the surface were sampled by time averages of the number of water molecules in cylindrical slabs parallel to the surface with a radius of 1.5 nm and height of 0.1 nm at equally distributed locations. The Fig. 4.4 shows density profiles in different locations for different humidity in the imidazole system. Under all conditions, the water density is zero at a distance of about 0.0 nm and 3.3 nm from the first silicon layer of the surface, which is taken as the point of origin. The range accessible to water molecules is about 3.3 nm, which corresponds approximatively to the pore diameter of 3.0 nm. In the middle of the pore, the water density is maximal with about 1 g cm^{-3} under full hydration.

4.3. DIFFERENT HUMIDITY

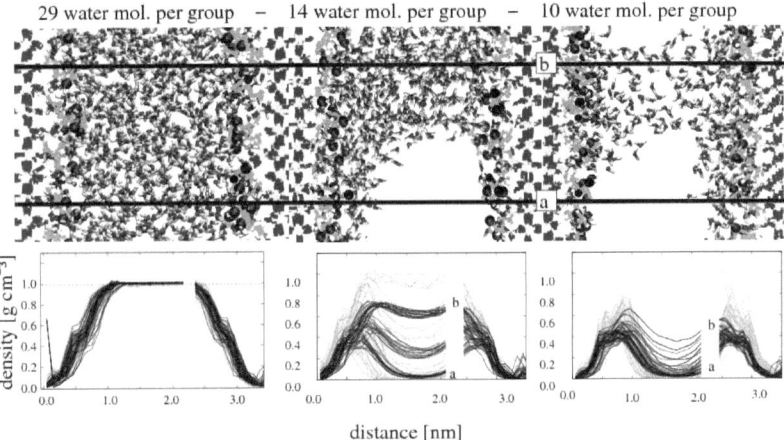

Figure 4.4: *Distributions of water molecules at different positions along the pore wall in imidazole functionalised pore at reduced water content, i.e. $\lambda =29$, $\lambda =14$, $\lambda =10$. Two temperatures are shown: 300 K - grey line, 450 K - black lines. The side views of the slab model illustrate the hydration at 300 K. To emphasise inhomogeneity, regions with low and high density are marked (a) and (b), respectively.*

This value - similar to bulk water - is reached already at about distance of 1.0 nm from the surface and symmetrically decreases at about 2.3 nm, which refers to a distance of about 1.0 nm from the other surface. We attribute the shoulders visible in the density profiles to the effective length of the functionalised groups, which corresponds also to the distances measured from the coordinates as assigned in Fig. 4.1. At larger distances, no direct interaction of water with the functional surface occurs. In the case of lower hydrations, regions with almost vanishing density are observed, which are connected to inhomogeneities in the water distribution. Therefore, an exclusively water based proton transport mechanism is expected to break down due to the lack of free water percolation paths going below at least $\lambda=14$. An increase in temperature to 450 K and therefore in kinetic energy leads to a homogenesation of the water distribution. This means that the density range is narrowed along the middle of the pore apparent from the sample with $\lambda=14$ at 450 K compared to the respective sample at 300 K. Further, the water density close to the surface is lower leading to a higher average density in the middle of the pore, which means a homogeneisation of the water distribution perpendicular to the surface. However, the studied samples with lower humidity still show an inhomogeneous distribution at 450 K.

As shown in Fig. 4.5, a qualitatively similar picture is drawn in the case of sulphonic acid functionalisation concerning temperature and hydration dependance. Still, the water density of fully hydrated systems of sulphonic acid is significantly higher than the density of imidazole functionalised pores. At $\lambda=31$, the maximal water density is reached closer to both surfaces compared with the imidazole functionalised system - at 0.8 nm and 2.5 nm distance from the first Si-layer, and the hydrophilic character of the functionalisation leads to a higher maximal density at these positions, which slightly exceeds bulk density. The latter is then reached in the middle of the pore.

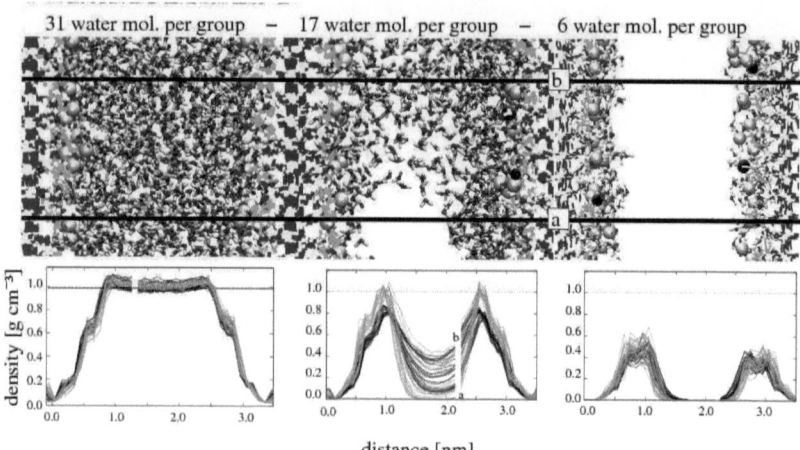

Figure 4.5: *Distributions of water molecules at different positions along the pore wall in sulphonic acid functionalised pore with reduced water content, i.e.* $\lambda = 31$, $\lambda = 17$, $\lambda = 6$. *Two temperatures are shown: 300 K - grey line, 450 K - black lines. The side views of the slab model illustrate the hydration at 300 K. To emphasise inhomogeneity, regions with low and high density are marked (a) and (b), respectively.*

4.3.2 Local Water Environment

Beside the density profile, the local water environment of the functional groups was characterised by RDF. The RDF's, see Sec. 1.2.1, were calculated between the proton conducting groups and the water molecules, see Sec. 3.2.2. The proton conducting groups were represented by the nitrogen atoms of imidazole or by the sulphur atoms of sulphonic acid. The water molecules were represented by their oxygen atoms. The volume integral over the RDF - normalised to the number of the representative atoms of the proton conducting group - results in the number density, which is defined as the average number of atoms inside a sphere around the proton conducting group, depending on the radius. Fig. 4.6 and Fig. 4.7 show the number density of water molecules in the vicinity of the functional groups. Despite the overall inhomogeneity of the system, a smooth increase is observed with increasing distance for all curves. Quantitatively, the average number of water molecules in a sphere of 1 nm³ volume (marked by a red dotted line in the Figure) varies from 8 to 18 molecules for the imidazole case with a total number of 29 to 5 water molecules per group (λ), which corresponds to 100 % to 17 % hydration samples at 450 K. In the sulphonic acid case for λ from 31 to 6 corresponds to 100 % to 19 % hydration samples, one obtains an average number of water molecules of 10 to 21. The density is higher in the proximity of sulphonic acid groups rather than of imidazole groups as expected, since the total water density at full hydration is also higher.

The inhomogeneity of the distribution, as mentioned above, leads to a broad variation of integrated RDF for arbitrarely chosen groups in models with the same hydration. For example, in the case of the imidazole system with 48 % hydration at 450 K ($\lambda = 14$), one can find a local environment for a single group that equals the ensemble averaged environment of the lowest or highest hydration case. On the chosen time scale of 1 ns, regions of homogeneity

4.3. DIFFERENT HUMIDITY

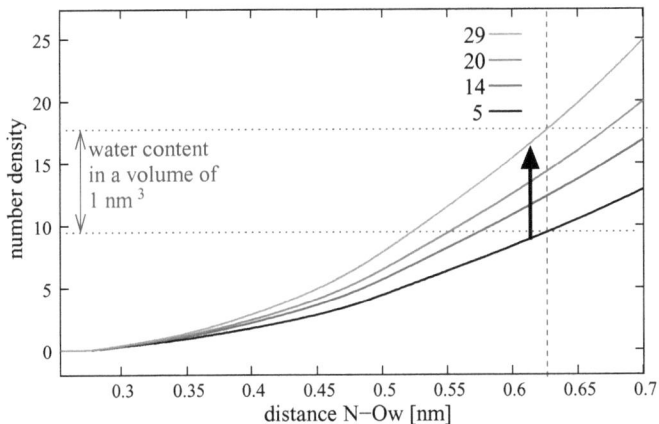

Figure 4.6: *Number density of water in the vicinity of the imidazole group for different values of λ, i.e. 5, 14, 20 and 29. Slab model functionalised with imidazole.*

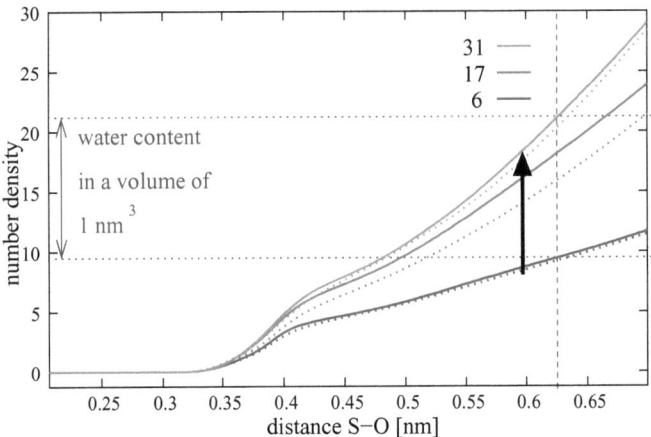

Figure 4.7: *Number density of water in the vicinity of the imidazole group for different values of λ, i.e. 6, 17 and 31. Slab model functionalised with (deprotonated) sulphonic acid.*

4.3.3 Interaction of Groups with the Surface

Beside the interaction with the water environment, the proton conducting groups also interact with the silicon dioxide surface. RDF's are calculated between proton conducting groups and the surface, the latter being represented by the silicon atoms. Especially at low hydration or even in the dry case, a direct interaction is expected. This means that all functional groups incline stronger towards the silanol surface in the dry case than in the hydrated case. From the integral over the RDF, i.e. the number density, the interaction of proton conducting groups with the silicon dioxide surface was estimated; when the number density reaches one, the radius refers to the average distance between the group and the first silicon atom. For comparison, a geometric estimate is calculated for a particle attached by an unflexible chain of length r to a flat surface, see Fig. 4.8. Under the condition that the particle visits all possible positions with the same probability, the average distance of the particle from the surface ($r\cos(x)$) evaluated as $\frac{4r}{3\pi}$. According to Fig. 4.1, the alkyl chain length is about (0.7 ± 0.1) nm and the distance between the base point and the first silicon layer equals (0.2 ± 0.05) nm; thus, the average distance amounts to (0.5 ± 0.1) nm.

The number density of silicon atoms[i] in the vicinity of nitrogen atoms (imidazole) or oxygen atoms (sulphonic acid), reaches 1 group at a distance of about 0.45 nm in the dry case for both systems. This average distance between groups and surface corresponds well with the lower range of the simple geometrical estimate, considering the lager error interval. Under hydration, the average distance to the surface is shifted to higher values by about 0.1 nm for imidazole system and 0.13 nm for sulphonic acid systems. In the case of sulphonic acid, the hydrated system is in the deprotonated state, while the dry system contains only neutral sulphonic acid groups. The average functional groups are inclined towards the surface due to the interaction with the hydroxyl groups. This interaction is shielded in the presence of water while the effect of temperature in both systems is negligible with a shift of about 0.01 nm. Going from the completely dry to a fully hydrated system, the distance to the surface decreases. The decrease is significantly stronger for sulphonic acid system, especially as the protonation state is changed and ionic sulphonic acid groups are considered under full hydration.

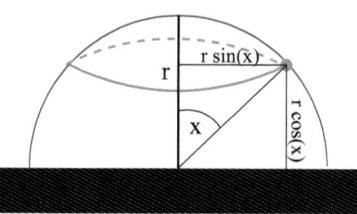

Figure 4.8: *Schematic view: Area that can be reached by a particle with a unflexible chain length r.*

4.4 Short Conclusion

The sulphonic acid functionalised silicon dioxide pores are more hydrophilic than the imidazole functionalised system. At full hydration, the number of water molecules per sulphonic acid group equals $\lambda = 31$ and per imidazole group equals $\lambda = 29$ at 450 K for the model

---

[i]For the RDF (Si-N and Si-O), silicon atoms directly bound to the alkyl chain were not taken into consideration for the analysis.

4.4. SHORT CONCLUSION

with a surface density of 1.3 nm^{-2} groups. At lower hydration, the distribution of water is inhomogeneous and the surface of the pore is covered by a water layer for all systems. By decreasing the hydration from 100% to 19% or 17%, the number of water molecules inside a spherical volume of 1 nm^3 around the functional group decreases from 22 to 9 in the case of sulphonic acid and from 18 to 8 in the case of imidazole. Beside the interaction with water, an interaction of the functional groups with the surface is observed, which is shielded under hydration.

Chapter 5

Proton Transport Inside the Porous Environment

In this chapter, proton transfer mechanisms are identified for the functionalised silicon dioxide model and the environmental influence is evaluated in comparison with functional molecules in vacuum, see Chapter 3. Three different proton transport mechanisms are expected: water based transport, transport between functional groups via water molecules and direct transport between functional groups via chain movement. The three different mechanisms are discussed for both functional groups, imidazole and sulphonic acid, and estimated based on the informations about the local environment inside the functionalised amorphous silicon dioxide pores from Chapter 4.

As Tuckerman et al. showed, the adiabatic approximation can be applied to the proton transport in water at room temperature and the atoms behave in an essentially classical manner[65]. Therefore, the proton and all other nuclei are treated as classical particles, while the electronic part is described quantum mechanically. In order to compare the different mechanisms, free energy barriers for simple model systems are calculated, computational details are given in Sec. 5.1. Moreover, the FF simulations of type (II) as explained previously (see Sec. 4.1.2) are evaluated according to the properties related to the proton transport mechanism and compared to results from Sec. 3.2 for functional molecules in vacuum.

5.1 Free Energy Barrier Calculation and Reaction Coordinate

Free energy barriers were obtained by WHAM from umbrella sampling simulations, as described in Sec. 1.3.2 and Sec. 1.3.3. All simulations were performed in a NVT ensemble at a temperature of 450 K using a Nosé-Hoover thermostat with a time constant of about 0.1 ps. Two models consist of isolated molecules of methyl sulphonic acid or methyl imidazole and water, as also used in Sec. 3.1. These simulations were performed with the CHARMM program and the entire system was described by H-bond-DFTB with the pbc/mio SK-file set. The third mechanism directly involves the immobilisation of the functional groups to the silicon dioxide substrate; therefore, a functionalised cristobalite slab model was used and the umbrella sampling was performed with the GROMACS program using the QM/MM coupling scheme. For this issue, reaction coordinates depending on mCEC were implemented in the GROMACS program, see Appendix B.

All reaction coordinates depend on the proton position, which is described by the mCEC, Sec. 1.4. The parameters of the mCEC were chosen in the following way. The nitrogen atoms

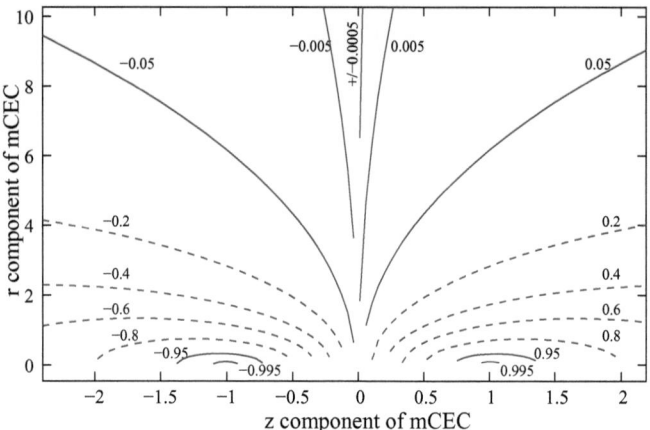

Figure 5.1: *Surfaces of constant ratio reaction coordinate: $D_0 \in [\pm 0.2, \pm 0.4, \pm 0.6, \pm 0.8]$ - dotted lines and $D_0 \in [\pm 0.0005, \pm 0.005, \pm 0.05, \pm 0.95, \pm 0.995]$ - straight line; the surface is parametrised by the relative mCEC in cylindrical coordinates; the reference groups are located at ± 1*

were weighted with (-0.5), the sulphonic acid group was not considered for the coordinate, the oxygen of the water molecules were weighted with (-2) and the hydrogen atoms of the water molecules as well as the hydrogen atoms bonded to the nitrogen were weighted with (+1). All other atoms are not considered for the coordinate.

In this work, two kinds of one dimensional reaction coordinate were applied, one describing the distance between a reference molecule (X_A) - proton conducting group - and the proton coordinate (mCEC), the other one is dimensionless and consists of the ratio between the distance of the mCEC from two different reference molecules (X_A and X_B). These reaction coordinates are called distance reaction coordinate and ratio reaction coordinate in the following.

Distance Reaction Coordinate

The exact functional form of the distance reaction coordinate ζ^d is given by the following, where $\vec{c}$ means the mCEC and $\vec{R}^A$ means the coordinates of the reference molecule (X_A):

$$\zeta^d = \left|\vec{c} - \vec{R}^A\right| = \sqrt{\sum_{m=1}^{3}(c_m - R_m^A)^2} \qquad (5.1)$$

The distance reaction coordinate ζ^d has the unit of a length. In the space of the mCEC, surfaces with constant ζ^d are spheres around the reference molecule X_A. This leads to the density of states $\Omega = 4\pi r^2 dr = 4\pi (\zeta^d)^2 d\zeta^d$. The free energy $F(\zeta^d)$ was proportional to $ln(\zeta^d)$ (see Formula 1.83), if the probability distribution ρ of the mCEC is homogeneous in space, i.e. the potential energy was constant.

5.2. WATER BASED PROTON TRANSPORT

Ratio Reaction Coordinate

The ratio reaction coordinate ζ^r is given by the following, where $\vec{c}$ means the mCEC and $\vec{R}^A$ and $\vec{R}^B$ mean the coordinates of the reference molecules (X_A and X_B):

$$\zeta^r = \frac{\zeta_A^d - \zeta_B^d}{\zeta_A^d + \zeta_B^d} \tag{5.2}$$

$$\zeta_{A/B}^d = \left|\vec{c} - \vec{R}^{A/B}\right| = \sqrt{\sum_{m=1}^{3}(c_m - R_m^{A/B})^2} \tag{5.3}$$

The ratio reaction coordinate ζ^r is dimensionless. To illustrate the ζ^r, cylindrical coordinates are introduced. The z-axis is in the direction of the vector between X^A and X^B. The radial coordinate r and the coordinate z are expressed as dimensionless coordinates referred to the distance [i]. The reaction coordinate is defined on the interval between -1 and 1. Fig. 5.1 shows the surface with constant reaction coordinate ($\zeta^r = D_0$) on this interval.

The functional form of the surface is given as $r(z)$ with a parameter range of z limited as denoted below.[ii]

$$r = \sqrt{\frac{2z}{D_0} - 1 - z^2}$$
$$D_0 \in \,]-1, 1[$$
$$z \in [Z_0, Z_1] = \left[\left(-\frac{1}{|D_0|} - \sqrt{\frac{1}{|D_0|^2} - 1}\right), \left(\frac{1}{|D_0|} + \sqrt{\frac{1}{|D_0|^2} - 1}\right)\right]$$

The density of states Ω show an asymptotic behaviour at $\zeta^r = 0$, since the surface of constant ratio reaction coordinate is not finite but rather an infinite plane. In the computer simulation, space and simulation time are finite.

5.2 Water Based Proton transport

As discussed in Sec. 2.1.1, water based proton transport is quite well studied in the literature. Being the preliminary event to water based transport, deprotonation of functional groups is discussed for the low water regime followed by the evaluation of self diffusion coefficients of hydronium ion in sulphonic acid functionalised silicon dioxide pores.

In Sec. 3.1, the proton affinities were calculated. Comparing protonation of imidazole, sulphonic acid, deprotonated sulphonic acid and various water clusters in vacuum, the energetically most favourable condition is charge neutrality of sulphonic acid. Furthermore, protonation of large water clusters of size bigger than 10 molecules energetically best a protonation of imidazole taking only potential energy differences into consideration.

Though, the relative deprotonation of methyl sulphonic acid groups in solution is known from spectroscopy to exceed 90% already at low water contents ($\lambda =7$)[81], but the protonated state of imidazole in aqueous solution is favoured by about 9 kcal/mol[77], as discussed in Sec. 2.1.2. These findings reveal clearly the importance of entropic effects and of the chemical environment in determining the chemical equilibrium, as they can not be explained by

---

[i]The coordinates of the space vector $\vec{x}$ are evaluated as $z = \frac{2\vec{x}\cdot\vec{r}_{AB}}{r_{AB}^2}$ and $r = \frac{2\sqrt{r_{AB}^2 x^2 - (\vec{x}\cdot\vec{r}_{AB})^2}}{r_{AB}^2}$; the angular cylindrical coordinate is not used, due to the symmetry.
[ii]$\zeta^r = 1$ if ($\zeta_B^d = 0$) or $\zeta^r = -1$ if ($\zeta_A^d = 0$). This means ($\vec{c} = \vec{R}^B$) or ($\vec{c} = \vec{R}^A$).

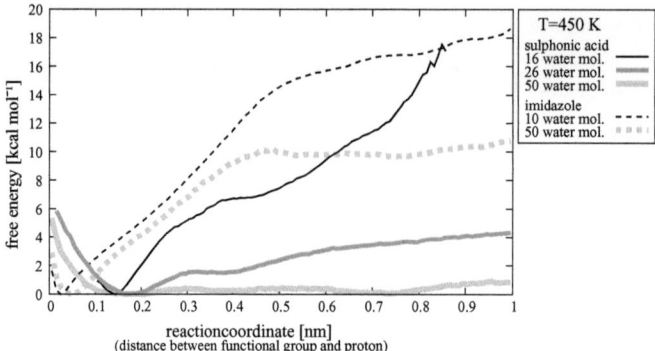

Figure 5.2: *FE barrier of deprotonation; methyl-sulphonic acid straight line, (protonated) methyl-imidazole dashed line; shift of minimum between both systems due to reference coordinate, S atom and center between the nitrogen atoms atom respectively*

the potential energy differences between the isolated molecules in vacuum.

As the mentioned measurements describe deprotonation in aqueous solutions, only the limiting case of fully hydrated systems is comparable to low concentrated solutions. In the interesting case of low humidity in a porous environment, water clusters are formed in the vicinity of the proton conducting species, as described in Sec. 4.1.2, and the density of groups is relatively low compared to aqueous solution with high concentration, i.e. low water content. Therefore, the free energy barriers were calculated for isolated methyl sulphonic acid or methyl imidazole molecules in a water environment that corresponds to the low humidity regime as characterised in Sec. 4.1.2.

5.2.1 Free Energy of Deprotonation

The deprotonation at 450 K of (neutral) methyl sulphonic acid and protonated methyl imidazole was studied. The systems consisted of different amounts of water molecules, which were constrained via hard wall boundary conditions to a sphere of 1.2 nm radius around one (isolated) proton conducting group. The groups were (neutral) methyl sulphonic acid and protonated methyl imidazole. The reaction coordinate was the distance between the proton conducting group and the proton coordinate mCEC, see Sec. 1.4. The group was represented by either the sulphur atom or the center of the nitrogen atoms. Each of the 20 windows of umbrella sampling were equilibrated for about 6 ps and the length of the trajectories was at least 2 ps.

Figure 5.2 shows the profile of the free energy along the reaction coordinate. The reference points in the two systems are different, thus the minimum is slightly shifted. Both minima signify the equilibrium position at the molecule. A further shift and a broadening of the minima is observed as the water cluster is increased for both systems. This corresponds to a shift of the equilibrium position of the proton in the cluster, as reported in the literature in the presence of 3 water molecules[84][85][86][87], see Sec. 2.1.2. This analysis is focused on segregation of the proton from the acid/base on a larger length scale.

For methyl sulphonic acid in the presence of 10 and 26 water molecules, a local minimum

5.2. WATER BASED PROTON TRANSPORT

in the free energy profile at about 0.4 nm from the sulphur atom is observed, which marks the location of the first solvation shell. The energy barrier for deprotonation of sulphonic acid vanishes completely in the presence of 50 water molecules, while in the presence of 26 water molecules a barrier of about 2 kcal/mol has to be overcome for deprotonation which agrees well with recently published results for fluorinated sulphonic acid[88], see Sec. 2.1.2. Even though the deprotonation involves charge separation[i], a complete deprotonation of sulphonic acid occurs at a water density with respect to the hard wall boundary conditions that equals about the density of a 20% hydrated system, see Sec. 4.1.2. The screening effect of water in the system and a strong influence of entropic effects lead to the acidic behaviour of the methyl sulphonic acid in a water environment of more than 26 water molecules.

Contrasting this, the deprotonation of the protonated methyl imidazole shows no qualitative change by increasing the number of water molecules up to 50. The comparison of proton affinities of the ideal water cluster with the methyl imidazole molecule suggested that the formation of protonated water clusters with more than 10 water molecules would be favourable compared to the protonation of an isolated imidazole molecule. Neither the consideration of an isolated imidazole molecule nor the ideal ordering of the water cluster, though, is realistic in a humid environment at elevated temperatures. The free energy calculation showed that the interaction with the water surrounding the protonated methyl imidazole has a significant influence on the stability of the configuration. The protonated state of imidazole is still favoured by about 10 kcal/mol, which agrees well with the free energy estimate made from measurements in aqueous solutions of about 9 kcal/mol[77] and the energy difference of 8 to 9 kcal/mol between the protonated and deprotonated state of the imidazole containing the amino acid histidine in aqueous environments obtained by recently published simulations [80][79], as already mentioned in Sec. 4.1.2.

5.2.2 Vehicular Diffusion Coefficient in Sulphonic Acid System

As the deprotonated state of sulphonic acid was confirmed for hydrations of more than about 20% at 450 K, the vehicular diffusion coefficient is evaluated from FF MD simulations at different humidity. For λ values of 6, 17 and 31[ii], the deprotonation limit with respect to the water density is reached in all three cases, whereas the number of water molecules per group is lower than 26 molecules for $\lambda = 6$ and $\lambda = 17$.

For deprotonated sulphonic acid systems, the diffusion coefficients of hydronium ions are

Table 5.1: *(intrinsic) hydronium ion diffusion in sulphonic acid functionalised slab model*

water	diffusion constant$[10^{-5} cm^2 s^{-1}]$		E_A
λ	T= 300 K	T= 450 K	[kcal/mol]
hydronium ions			
6	0.03 ± 0.01	0.11 ± 0.01	2.4 ± 0.8
17	0.267 ± 0.003	0.9 ± 0.2	2.1 ± 0.4
31	0.56 ± 0.04	2.0 ± 0.3	2.3 ± 0.4
water molecules			
6	0.68 ± 0.09	9.8 ± 0.6	4.8 ± 0.3
17	1.7 ± 0.1	10.6 ± 0.9	3.2 ± 0.3
31	2.5 ± 1.4	8.66 ± 0.09	2.2 ± 0.1

---

[i]The charge separation leads to the high value of proton affinity for the deprotonated sulphonic acid.
[ii]corresponding to 19%, 55% and 100%hydration at 450 K, see Chapter 4

calculated from the data of simulations of type (II), see Sec. 4.2.1. Therefore, the mean squared displacement for the hydronium ion was calculated from the MD trajectories for the adequate functionalised system, as described in Sec. 4.1. Data from the previously discussed NVT simulation (see Sec. 4.1.2) was considered for a humidity with λ ranging from 6 to 31, which refers to a humidity of 23%, 46% and 95% compared to full hydration at 450 K. As vehicular diffusion in water becomes more important than structural diffusion at high temperatures, vehicular (hydronium ion) self diffusion is a good approximation for total protonic self diffusion, especially in the case of low hydration. The activation energies were calculated for the diffusion at different hydration levels considering an Arrhenius-like behaviour. For a small temperature range the temperature dependance of the prefactor is negligible. In the case of FF MD, one can exclude the appearance of different processes due to the classical description of the particles by predefined parameters. The limited description of the FF thereby justifies the application of Arrhenius on a larger temperature scale. The diffusion constant was calculated for two different temperatures. The error was estimated only by error propagation.

In Tab. 5.1, a strong dependance of the diffusion constant on temperature and water content is observed. Especially for low water content, proton diffusion is low, as the water layer is not sufficient to screen the electrostatic interaction. The water diffusion is increased with increasing water content due to interaction with the surface, but this effect is less pronounced. Especially at low hydration, a strong effect of temperature on the water diffusion is observed, as the mobility of the water molecules is increased. This is also reflected in a change of the activation energy.

Nafion is the standard reference material for PEM. According to experimental studies, the diffusion coefficient of Nafion is in the range of $0.5\text{-}2\cdot10^{-5}cm^2s^{-1}$ and activation energies are in the range of 3 to 5 kcal/mol[89][90][91]. The activation energies from MD simulations are reported to be 4 kcal/mol for water and about 2 kcal/mol for the hydronium ion at different λ[94][95]. These results compare well with the activation energies for hydronium diffusion inside a functionalised particle (about 2.5 kcal/mol) and for water (2 - 5 kcal/mol) that are reported here. From a classical MD simulation for a Nafion system with a water sulphonic acid ratio of λ=14 at 300 K, the diffusion coefficient for water equals $0.6\cdot10^{-5}cm^2s^{-1}$, while the diffusion coefficient for the hydronium ion is lower by a factor of 3 ($0.2\cdot10^{-5}cm^2s^{-1}$)[94]. These diffusion coefficients are similar to diffusion in the pore at the same temperature, suggesting a good proton conduction inside the pore environment equivalent to the proton conduction in pure Nafion. In the combined system (particles in Nafion), both regions are expected to participate directly in the proton transport, as similar proton conducting ability means equivalent pathes through the material. The functionalised particle has an active proton conducting function and does not only influence the chemical environment of Nafion. This argumentation does not take into account boundary effects in the hybrid material and the possibility of inhomogeneous water distribution at low humidity. On the other side, direct participation of the particles in the proton transport is also supported by experimental conductivity measurements of mixed Nafion/functionalized-SiO_2 systems, which show even higher conductivity at higher temperatures [104].

5.3 Proton Transport involving Functional Groups

The water dominated mechanism described in the previous section is only suitable for sulphonic acid systems in humidity conditions of more than 20% (with respect to full hydration at 450 K). In low hydration conditions and in the imidazole case, instead, the proton conduct-

5.3. PROTON TRANSPORT INVOLVING FUNCTIONAL GROUPS

ing species are expected to be directly involved in the proton transport. To clarify the proton transport mechanisms at these conditions, the approach of Sec 3.2 is followed. The number of isolated groups and the hydrogen bond fluctuations were identified as useful indecators for the direct proton transport for the isolated immobilised functional molecules in vacuum and in this section, the influence of the environment is discussed. Subsequently, free energy barriers are calculated for the proton transport between functional groups, depending either on the local water environment or on the mobility of the functional molecules.

5.3.1 Direct Transport between two Groups

The distribution of proton conducting groups was analysed similar to the approach described in Sec. 3.2. Therefore, the number density in a sphere around the functional group and the distribution of aggregates with size N was calculated from MD simulations of type (II), see Sec. 4.1.2. The aggregate consists of functional groups with a distance of less than 0.35 nm, measured from the nitrogen or oxygen position, respectively. The slab model for amorphous silicon dioxide is functionalised with pentyl sulphonic acid or imidazole using a surface density of 1.3 nm^{-2}. Different temperature and humidity conditions are considered.

Table 5.2: *Percentages and stability of isolated groups under the influence of the environment at different water contents and functionalisations.*

lambda	water content [%]	Temp. [K]	isolated groups [%]	stability after 25 ps [%]	inactive groups [%]
functionalisation (1.3/nm^2): pentyl imidazole					
0	0	300	81± 3	97 ± 5	79±7
0	0	450	97± 3	85 ± 5	82±7
5	14	300	86± 3	87± 5	75±7
5	17	450	80± 3	77± 5	62±6
10	28	300	84± 3	83± 5	70±7
10	34	450	83± 3	73± 5	61±6
14	39	300	86± 3	85± 5	73±7
14	48	450	83± 3	74± 5	61±6
19	56	300	85± 3	81± 5	69±7
19	69	450	83± 3	72± 5	60±6
29	81	300	86± 3	90± 5	77±7
29	100	450	84± 3	69± 5	58±6
functionalisation (1.3/nm^2): neutral pentyl sulphonic acid (SO$_3$H)					
0	0	300	73± 3	88± 5	64±6
0	0	450	69± 3	91± 5	62±6
15	38	300	90± 3	87± 5	78±7
15	41	450	86± 3	82± 5	71±7
functionalisation (1.3/nm^2): deprotonated pentyl sulphonic acid (SO$_3^-$)					
6	16	300	84± 3	-	
6	19	450	80± 3	-	
17	46	300	89± 3	-	
17	55	450	85± 3	-	
31	84	300	89± 3	-	
31	100	450	90± 3	-	

RDF and Aggregates

For both systems, imidazole and sulphonic acid, the number density becomes smoother and it is decreased for small distances comparing the functionalised slab model with the analogous system of isolated functional groups (B/1.0 nm) of Sec. 3.2. The integrated RDF hints to a more homogeneous distribution of proton conducting groups due to the influence of the chemical environment. In Fig. 5.3, the number densities of sulphonic acid groups are shown. The increase in water content (indicated by the black arrow) and temperature lead to a further shift in the average distance between the sulphonic acid groups, from about 0.48 nm to a range of 0.55 nm to 0.73 nm. To evaluate this effect quantitatively, Tab. 5.2 lists the percentages of isolated groups for both systems, imidazole and sulphonic acid. The isolated groups corresponds to aggregates of size $N = 1$.

In the dry case ($\lambda = 0$), the percentage of isolated imidazole molecules is higher than the one of neutral sulphonic acid molecules. As in model (B/1.0 nm), it is noticed that (neutral) sulphonic acid has a significant higher probability to aggregate ($\approx$30%) compared to imidazole (10-20%) under the influence of the chemical environment at dry conditions. The probability is higher by a factor of more than two at 450 K. For both systems, imidazole and sulphonic acid, the dry case is special as the functional groups only interact with the hydroxyl groups on the surface. This interaction reduces the probability of aggregation significantly, as the number of isolated groups is approximately doubled compared to the results of Sec. 3.2.

In the case of imidazole, the percentage of single molecules is about 83% at 450 K and 86% at 300 K, averaging the results of all systems with some humidity. Humidity dependence is only observed in the limit of dry systems where the percentage of isolated groups is increased by about 10% at 450 K and decreased by about 5% at 300 K going from hydration to the dry case, while the deviations from the average are lower than the error estimate for all other water contents. This can be interpreted as a strong influence of the first solvation shells on the interaction between the groups. A further increase in the number of water molecules has no effect on the interaction.

In the deprotonated sulphonic acid system, the average percentage of single molecules is slightly higher than in the imidazole case, with up to about 90%. For sulphonic acid, the dependence on the water content is even more pronounced, as a change in the protonation state is expected, but differences between the neutral system for $\lambda = 15$ and the deprotonated system for $\lambda = 17$ are minor. The percentage of isolated groups nevertheless increases constantly with increasing water content, going from about 70% to about 90% of isolated groups. In the picture of deprotonated sulphonic acid molecules, water based proton transfer is dominant under more than about 20% hydration, as discussed in Sec. 5.2. From trajectory observation, aggregation of deprotonated groups is known to coincide with the aggregation of at least one hydronium ion and is interpreted as a hindered dynamic acid-base-equilibrium due to the FF description. In the picture of deprotonation, RDF and aggregates distribution are not the relevant indicators, as the mechanism is sufficiently described by vehicular diffusion, see Sec. 5.2.

Fluctuations of Aggregates

For the imidazole system and the neutral sulphonic acid system at 450 K, fluctuations of isolated groups are evaluated. In the dry system, about 90% of the isolated groups are isolated for more than 25 ps considering the error estimate. The fluctuations of the neutral sulphonic acid system are approximately independent of temperature. This hints to a strong hydrogen bond network. For imidazole the temperature effect is more pronounced. In humidity

5.3. PROTON TRANSPORT INVOLVING FUNCTIONAL GROUPS

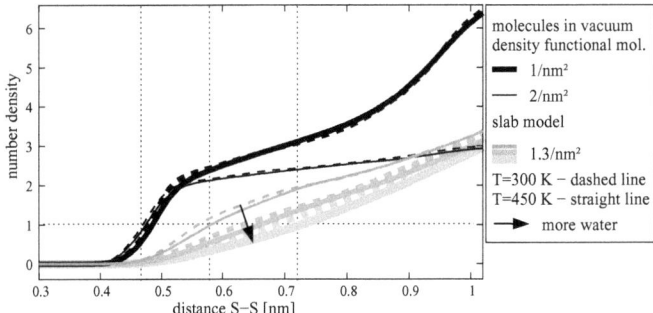

Figure 5.3: *Number density obtained by integration of RDF; comparison of model B/1.0 nm, B/0.7 nm and the deprotonated sulphonic acid functionalized slab model with density of 1.3 groups per nm^2 for different water contents and different temperatures. The water content is listed in Tab. 5.2*

conditions at 450 K, the stability of isolated imidazole groups is significantly lower than for sulphonic acid.

It is possible to define a measure, called percentage of inactive groups for the direct proton transport, which corresponds the percentage of isolated groups that stay isolated during at least 25 ps and thus do not participate in proton transport on a reasonable time scale. As most aggregates in the dry case are formed in the sulphonic acid system, the number of inactive groups is low. A similar low value is achieved for imidazole under hydration at 450 K resulting form increased fluctuations. At 450 K under hydration, the fractions are significantly lower for imidazole, as even both quantities, the percentage of isolated groups as well as the stability decreases with temperature; while for sulphic acid the temperature effect is negligible. Comparing to the results from Sec. 3.2, the influence of the environment is expected to lower direct proton transport, as the number of inactive groups is below 15% for all functional groups in vacuum, or even vanishes (Tab. 3.3). The average distances between proton conducting groups are increased under the influence of the environment and the percentage of inactive groups is increased. In the following, the importance of both mechanisms should be clarified for the imidazole case.

5.3.2 FE of Water based Transport between two Groups

The average distances between proton conducting groups are increased under the influence of the environment, as the previous section showed, and the percentages of inactive groups exceed 50% in both systems. Thus, the direct proton transport is reduced significantly. Therefore, a direct participation of water molecules in the proton transport is described in the following.

Proton transport between functional groups is analysed in low humidity conditions corresponding to less than 16 % and for two different distances between the functional groups, which can be identified with the average distance in a region of high or low density of functional molecules. A distance of 1.2 nm between the carbon atoms of the methyl groups is in agreement with the estimate for the 1.3 nm^{-2} sulphonic acid functionalised slab model. Under hydration, the average distance between the groups is shifted to distances of about 0.7 nm

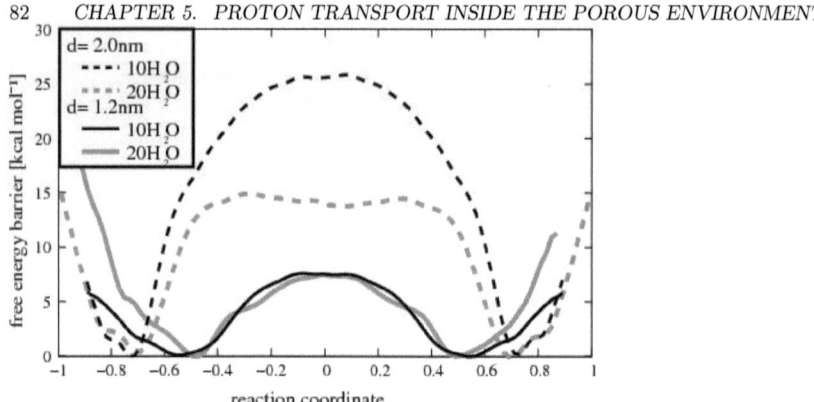

Figure 5.4: *Free energy barrier for the proton transport between two methyl sulphonic acid molecules depending on the water content and on the distance between the methyl groups*

between the sulphur atoms, see Sec. 5.3. The distance of 2.0 nm corresponds to an average distance between functional groups in system with lower surface density, i.e. 0.25 nm^{-2}. Taking into account an inhomogeneious distribution of functional molecules, regions of low density represent the bottleneck of proton transport. In Sec. 3.2.3, it was shown that at a distance of 2.0 nm between the functional molecules, the distance between groups corresponds to the periodicity of the system, as the groups are non directly interacting. This case is considered to represent a limiting situation.

The free energy barriers for proton transport were calculated as explained in Sec. 5.1. The system consisted of two proton conducting groups, methyl sulphonic acid groups or methyl-imidazole groups, constrained at a defined distance of 1.2 nm to 2.0 nm by fixing the methyl-carbon positions. A number of 10 to 20 water molecules were added to the simulation box. The water molecules were again constained by hard wall boundary conditions in a sphere of 1.5 nm radius around the groups. In the imidazole simulation, an excess proton was taken into account, while the sulphonic acid system had a proton deficit. As a reaction coordinate, the relative distance between the proton conducting species and the mCEC is chosen, as explained in detail in Sec. 1.4. The proton conducting groups were chosen as reference groups, assigned by sulphur atoms or nitrogen atoms. Each of the 20 windows of umbrella sampling were equilibrated for at least 6 ps and the length of the trajectory was at least 2 ps. As the histograms are of Gaussian like shape and show good overlap, the data set is believed to sample the phase space sufficiently.

Fig. 5.4 and Fig. 5.5 show the free energy barrier for the proton transport from one proton conducting group to the other. Their positions are assigned by the minima in the free energy graph. The abscissa corresponds to the relative reaction coordinate, which is defined as the relative distance between the proton coordinate mCEC and the reference groups at the proton conducting molecules included in each simulation. There is a decrease in the barrier with increasing water content from 10 to 20 molecules for systems with a group distance of 2.0 nm. In the sulphonic acid system, the decrease is more pronounced than in the imidazole case; from 26 to 15 and from 17 to 13 kcal/mol, respectively. A similar decrease happens for the imidazole by reducing the distance between the groups in the presence of 10 water molecules. In the sulphonic system, the decrease in the barrier is more drastical, i.e. about 8 kcal/mol

5.3. PROTON TRANSPORT INVOLVING FUNCTIONAL GROUPS

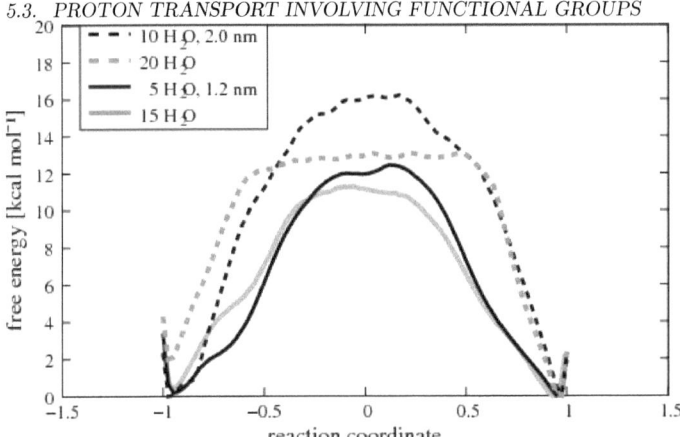

Figure 5.5: *Free energy barrier for the proton transport between two methyl imidazole molecules depending on the water content and on the distance between the methyl groups*

for methyl sulphonic acid at a distance of 1.2 nm for both water contents. For both systems, the variation in the water content has nearly no effect on the barrier for proton transport between groups at a short distance of 1.2 nm. In the imidazole case, the barrier limit of about 11 kcal/mol corresponds well with the free energy difference between protonated and deprotonated state of imidazole in the presence of 50 water molecules, see Sec. 5.2.1. One would expect the same for sulphonic acid, i.e. as the number of molecules is increased further, the barrier for the proton transport should go to zero. This comparison points out the strong effect of water in a sulphonic acid system, which is also expressed in the significant increase of the transport barrier between methyl sulphonic acid molecules at a distance of 1.2 nm.

5.3.3 Direct Proton Transport on the Substrate[1]

Previously, proton transport via water molecules from one proton conducting group to another was studied. Water based proton transport is considered to be the dominant mechanism in sulphonic acid systems. In imidazole systems, it instead plays a minor role, as the proton transport from an imidazole group to water has a barrier of more than 10 kcal/mol even in the presence of 50 water molecules. Previously, a percentage of 81-57% inactive groups was estimated at λ=0-22 evaluatingdirect proton transpor. To compare both proton transport mechanisms at the same level, free energy barriers were calculated for the direct transport between two imidazole functional molecules under environmental influence. Fig. 5.6 provides a schematic view of the different systems.

The transport between two functional groups via chain movement is described. The fully periodic system model consists of an α cristobalite slab, see Sec. 4.1. The surface area of the slab is about 10 nm^2 and it is functionalised with 3 or 6 pentyl imidazole groups seperated by 0.7 nm or 1.4 nm. They are arranged in a linear order as in model system A of Sec. 3.2.2, but here covalently bound to the surface. QM/MM coupling (Sec. 1.1.4) is used with a QM zone consisting of the methyl imidazole groups. The rest of the system is described by FF as explained above (see Sec. 4.1.2). Both model systems are considered in vacuum. In addition, the model with 6 functional molecules was studied under the influence of 100 water molecules.

84   CHAPTER 5. PROTON TRANSPORT INSIDE THE POROUS ENVIRONMENT

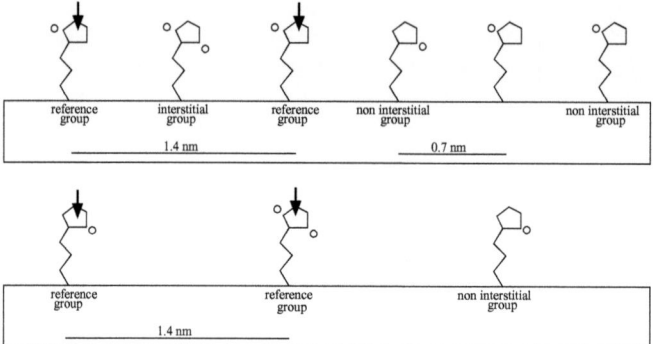

Figure 5.6: *Schematic view of model systems: upper part - model system with 6 functional imidazole molecules and distance d = 0.7 nm, below - model system with 3 functional imidazole molecules and distance d = 1.4 nm; large rectangle - silicondioxide; reference groups are marked by arrow; interstitial and non interstitial groups are assigned in the picture.*

The water - being part of the environment - was introduced in the MM zone, as transport via water molecules was explicitly excluded here and only the electrostatic influence and shielding effects of the water are studied.

Umbrella sampling was performed along the reaction coordinate describing the proton transport on the substrate. Two types of one dimensional reaction coordinates were specified: first, the distance coordinate - the distance between one proton conducting group and the proton coordinate mCEC; second, the ratio coordinate - the relative distance between mCEC and two proton conducting groups. Compared to the previously described umbrella simulations, a longer sampling is needed as the model contains more complex molecules and the chain movement broadens the phase space significantly. The simulation lengths range from 30.0 ps to 50.0 ps and the number of trajectory ranges from 30 to 110, resulting in an amount of all data in the range of 0.5 ns to 4.5 ns, see Tab. 5.3. The data from simulations with the distance reaction coordinate were evaluated in two different ways. Firstly, the data was taken after an equilibration of at least 30 ps, secondly, the equilibration was done more carefully equilibrated, resulting in less data for the production run, as each window was equilibrated for at least 50 ps with exactly the same parameters for the umbrella sampling.

Fig. 5.7 and Fig. 5.8 show the free energy profile along the reaction coordinate (or order parameter). Two different types of reaction coordinates were chosen in order to improve the interpretation. One notices that both coordinates lead to similar results for the energy barrier as listed in Tab. 5.3. The energy barrier is about 5 to 6 kcal/mol for all simulations. For the distance reaction coordinate, the WHAM analysis was performed for two different sizes of the data set, in order to get an estimate for an statistical error.

Distance Reaction Coordinate

Interpreting the distance reaction coordinate, the reaction coordinate is equivalent to a real space coordinate. The maxima are at a distance of 0.2 nm for all models and the local minima are located at 0.0 nm and 0.49 nm. The first minimum referres to protonation of the reference

5.3. PROTON TRANSPORT INVOLVING FUNCTIONAL GROUPS

group number	trajectory number	[ns]	all data [ns]	energy [kcal/mol]
reaction coordinate: distance				
6	69	30	2.1	6± 2
6	27	30	0.8	6± 2
3	113	39	4.4	7 ± 2
3	32	36	1.2	6 ± 2
reaction coordinate: ratio				
6	38	30	0.6	5 ± 2
3	52	50	1.3	5 ± 2

Table 5.3: *FE for direct proton transport on substrate; silicon dioxide functionalised with imidazole groups; simulation at 450 K; in the table different reaction coordinates are listed.*

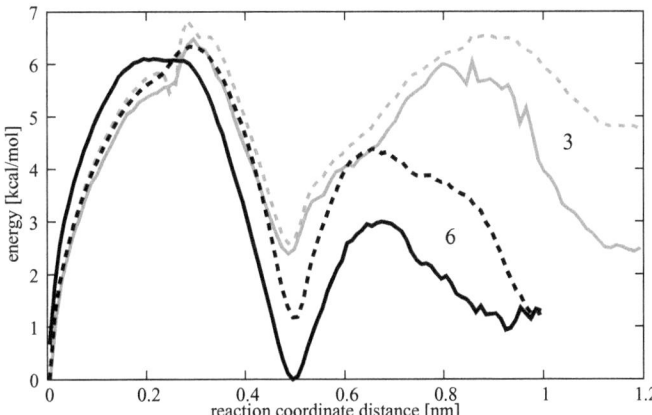

Figure 5.7: *Free energy barrier in kcal/mol for the proton transport on a imidazole functionalised silicon dioxide model at T=450 K: black line - d =0.7 nm, grey line - d =1.4 nm; dashed line - larger data set with less equilibrated data. The distance reaction coordinate is used.*

group. The second minimum marks the situation where two imidazole rings are aligned, one being the reference group, the other being protonated. The configuration is close to a Zundel configuration but is not recognised as such by the mCEC coordinate.[i] Furthermore, a second maximum is found at about 0.7 nm for the model with 6 groups and at 0.9 nm for the model with 3 groups. The difference between the two model systems, d=0.7 nm and d=1.4 nm, is expressed by the location of the next local minimum at about 1.0 nm or at about 1.2 nm.

In the range of the second maximum, the transfer of the proton will take place via a vehicle mechanism and is dependent on the chain movement of the adjacent functional molecule. The energy differences between the local minimum at about 0.5 nm and the second maximum are about 3 kcal/mol for all system. The chain movement is dominated by the MM zone (FF parameters) of the simulation. Therefore, an estimate for the chain movement is directly calculated from the FF-MD of Sec. 5.3. At the beginning of this section, radial distribution functions were discussed. Neglecting normalisation, N-N RDF corresponds to the histogram

---
[i] For a discussion of the mCEC parameters see below.

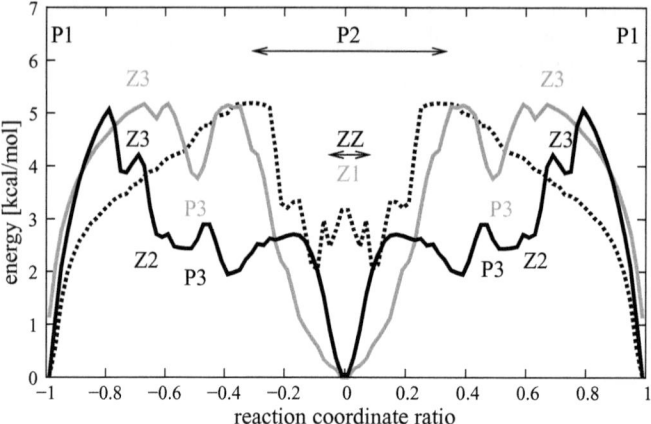

Figure 5.8: *Free energy barriers in kcal/mol for the proton transport on an imidazole functionalised silicon dioxide model at T=450 K: black line - d =0.7 nm, grey line - d =1.4 nm and dashed black line - d =0.7 nm in the presence of 100 water molecules (MM zone). The ratio reaction coordinate is used. See 5.9 for the notation of the configurations.*

over the reaction coordinate distance between two nitrogen atoms, e.g. the distribution function $\rho(\zeta_{NN-RDF})$. One obtains about 3.5 kcal/mol for the dry slab model at 300 K, while at higher temperatures or under hydration the barrier decreases to 1 kcal/mol, as the free energy curve is smoothed. These FF-barriers are significantly lower than free energy surfaces obtained from QM/MM umbrella sampling, since only the chain movement and aggregation is described by the FF. The quantum mechanical description of functional groups comprehend further mechanisms such as proton transfer between the groups and formation of a Zundel complex.

Ratio Reaction Coordinate

The ratio reaction coordinate was already applied in Sec. 5.3.2. It is symmetric and dimensionless. In Fig. 5.9, possible configurations for the imidazole systems are illustrated, three configurations with only one protonated molecule, P1, P2 and P3, and three different Zundel like configurations, Z1, Z2 and Z3. In Fig. 5.8, both minima at ±1 unambiguously refer to the protonation of a reference imidazole group (P1). This means that the reaction states are well defined. Interpreting other features is more difficult, as the denominator of the expression changes with the chain movement of the reference groups. If the reaction coordinate equals zero, the configuration is symmetric, with the mCEC exactly in the middle of the two reference groups. On the one hand this is realised by protonation of the interstitial imidazole group (P2), on the other hand by the formation of a Zundel ion (Z1) between the two reference groups. Configuration (P2) is only possible for the model with 6 groups and d =0.7 nm, as it involves an interstitial group, while configuration (Z1) belongs to the model with 3 groups and d =1.4 nm, see Fig. 5.6.

Considering an average distance of 1.4 nm between the reference groups, the formation of a Zundel complex from a reference group and an interstitial group (Z2) for the model with 6 functional groups results in a value of the reaction coordinate of ±0.65. For both models,

5.3. PROTON TRANSPORT INVOLVING FUNCTIONAL GROUPS

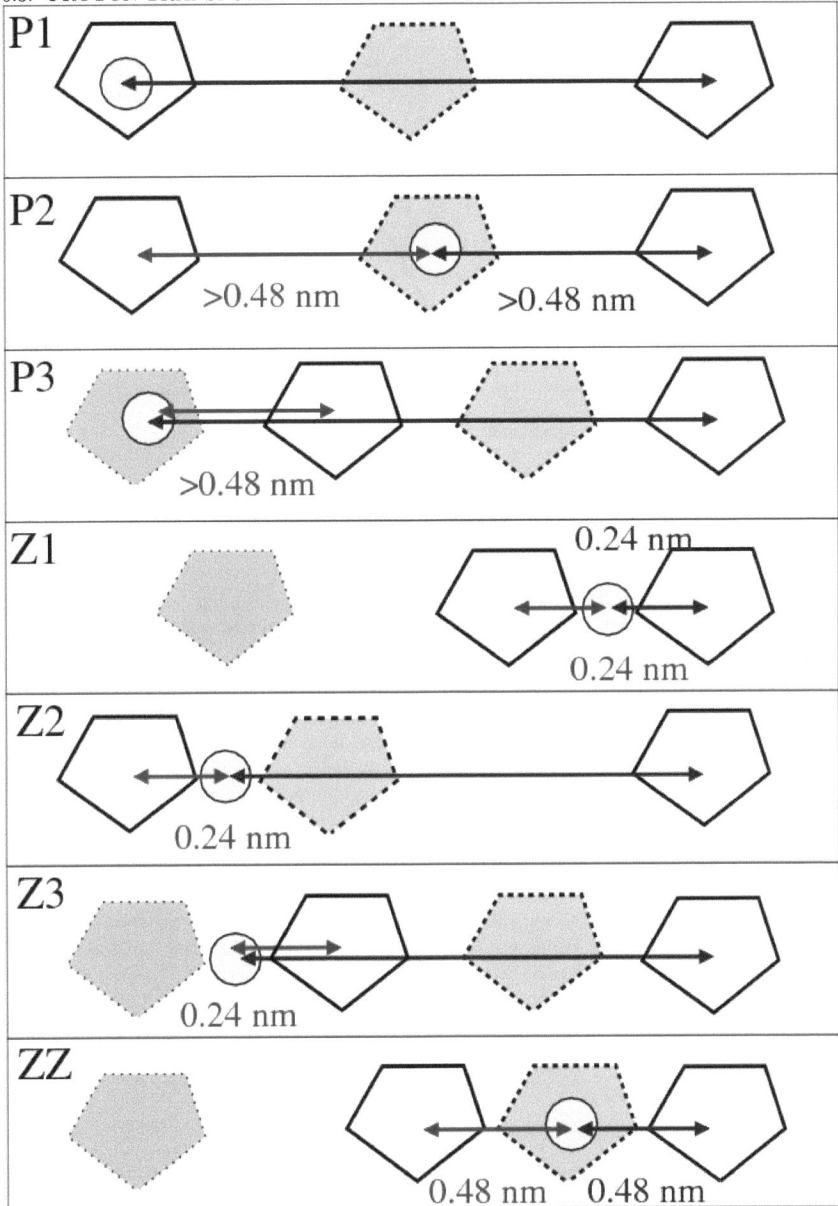

Figure 5.9: *Schematic view of configurations in the imidazole system described by the ratio reaction coordinate. The reference groups are drawn in solid black line, the interstitial groups - thick dashed black lines filled with dark grey; non interstitial - thin dotted black lines filled with grey; circle filled with light grey - mCEC; arrow - distance between reference group and mCEC.*

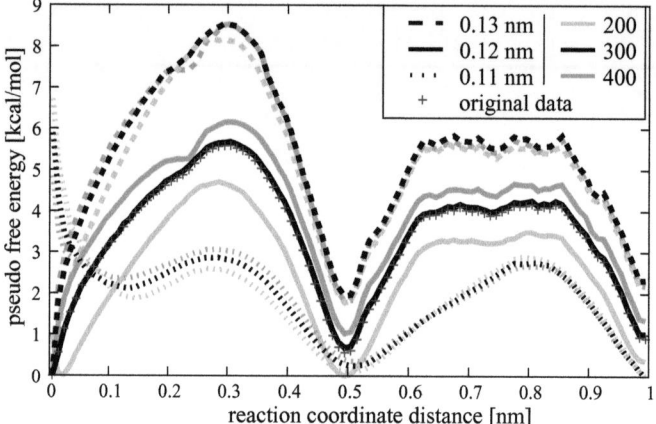

Figure 5.10: *Pseudo free energy barrier in kcal/mol obtained from histograms for the distance reaction coordinate depending on mCEC coordinate with changed parameters, K^0 and K^1, for the model with 6 groups. K^0 equals 0.13 nm (dashed), 0.12 nm (solid) or 0.11 nm (dotted); K^1 equals 200 (light grey), 300 (black) or 400 (dark grey). Free energy (k^0=0.12 nm and k^1=300) - crosses.*

the formation of a Zundel complex from a reference group and a non-interstitial group (Z3) is a possible configuration with an estimated value of the reaction coordinate of ±0.75.

The dry system model with 6 groups (d=0.7 nm) shows a strong maximum with 5 kcal/mol at ±0.8 and it drops by a factor of 2 at ±0.65. At this value, the configurations (Z3) and (Z2) contribute to the density of states. The sharp (global) minimum is located at 0.0, the region ±0.15 around the minimum shows parabolic behaviour. The broad minimum between -0.7 and 0.7 belongs to the Zundel like state (Z2) and to the protonation of the interstitial group at asymetric configurations (P2). The much sharper minimum at 0.0 refers to the symmetric alignment of all three groups (ZZ), reference groups and protonated interstitial molecule. Under the influence of FF-water, the formation of group aggregates is suppressed (as is already known from the RDF/aggregate-analysis of the FF MD; see above). For the free energy barrier with MM water, the broad minimum with about 3 kcal/mol, which mainly refers to configuration P2, is narrowed because of an increased distance between the reference groups. The Zundel like configurations, ZZ, Z2 and Z1, are weakened and the broadened minimum for the configurations (P1) could be interpreted as an effect of solvation. The dry system of model with 3 groups (d=1.4 nm), the minimum at a value of 0.0 (Z1) is broadened significantly due to a the decrease in the denominator to 0.48 nm; therefore, bond vibrations affecting the mCEC coordinate lead to significant changes in the value of the reaction coordinate.

Influence of the Choice of mCEC Parameters on the FE

The mCEC coordinate depends on the parameters $k^0 = 0.12$ nm and $k^1 = 300$, see Sec. 1.4. To esimate the parameter dependence of the mCEC, the following analysis was performed on the data from the umbrella sampling with the reaction coordinate $\zeta(k^1, k^0)$. The data of each sampling window was evaluated with the modified mCEC parameters K^1 and K^0. Then, a histogram of the reaction coordinate $\zeta(K^1, K^0)$ was calculated. These new histograms are

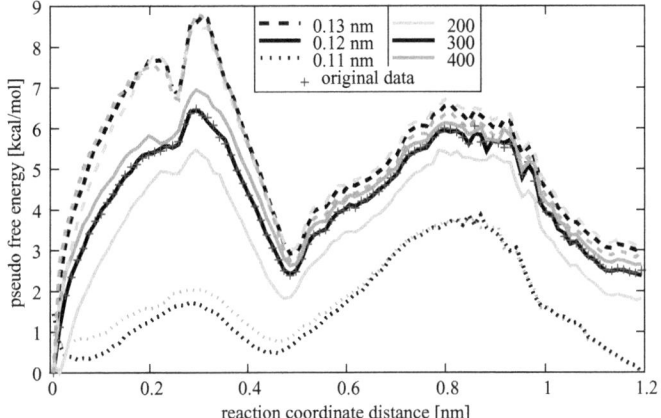

Figure 5.11: *Pseudo free energy barrier in kcal/mol obtained from histograms for the distance reaction coordinate depending on mCEC coordinate with changed parameters, K^0 and K^1, for the model with 3 groups. K^0 equals 0.13 nm (dashed), 0.12 nm (solid) or 0.11 nm (dotted); K^1 equals 200 (light grey), 300 (black) or 400 (dark grey). Free energy (k^0=0.12 nm and k^1=300) - crosses.*

interpreted using the WHAM algorithm, where the reference point for umbrella sampling is approximated by the reaction coordinate $\zeta(K^1, K^0)$, even though the umbrella potential was applied to $\zeta(k^1, k^0)$. The WHAM output is referred to as a pseudo free energy. Pseudo free energies approximate the real free energies if the deviation of the coordinates is negligible and the sampling is only slightly influenced by the changes in the parameters.

Fig. 5.10 and Fig. 5.11 show the pseudo free energies obtained from the new histograms via WHAM for the model of 6 and of 3 groups on the surface, respectively. The influence of the parameters of the θ-switching function is studied. The original parameters were k^0=0.12 nm and the steepness k^1=300, see Sec. 1.4. The Figure shows the pseudo free energy for the parameter k^0 ranging from 0.11 nm to 0.13 nm and the parameter k^1 from 200 to 400. For k^0=0.11 nm, the first maximum is significantly lower (especially for the system with 3 functional groups), which means that the probability distribution in this region is significantly increased. At the same time, the first minimum is shifted and the free energy is increased, which hints to a failure of the geometrical correction, as can be seen observing the coordinate with a molecular viewer. For k^0=0.12 nm, the switching function is lower than one for a bond distance (N-H) of 0.1 nm, see Fig. 1.3. For k^0=0.13 nm, the deviation is only about 2.5 kcal/mol, while the influence of the steepness parameter is even less. Here, the (pseudo) free energy only changes by less than 1 kcal/mol. The characteristics of the curve are similar.

5.4 Short Conclusion

Three possible proton conducting mechanisms have been analysed and compared for two different systems, imidazole functionalised and sulphonic acid functionalised silicon dioxide models. The direct transport was already discussed in Sec. 3 and was shown to consist of three steps; firstly, the aggregation of two or more functional groups, secondly, a barrier free proton transport between these groups, thirdly, the separation of the groups due to fluctuations in the hydrogen bond network and movement of the carbon chain. The other two proton transport mechanisms do not rely on the formation of aggregates and fluctuations, they directly involve proton transport in water. One possibility is the complete deprotonation of the functional groups, followed by water based proton transport as expected for acidic system. Another possibility is a water based proton transport over short distances from one proton conducting group to another. These three mechanisms are in direct competition. Under the influence of the chemical environment, the number of inactive groups, which do not participate in the direct proton transport on a time scale of 25 ps, is significantly increased and the average distance between two functional groups is elongated especially for sulphonic acid systems. This corresponds to a decrease in direct proton transport compared with the vacuum system discussed before.

In imidazole systems, the direct transport is nevertheless expected to be the dominating mechanism, as water plays a minor role. For a water dependent proton transport, the free energy barrier is at least by a factor of two higher than for the direct mechanism. In contrast to imidazole, deprotonation of sulphonic acid is expected from free energy calculations in a water environment of more than 20% humidity at 450 K inside a 3.0 nm pore. The formation of intrinsic charge carriers enables a water based proton transport, which was estimated by hydronium ion diffusion in FF simulations resulting in an activation energy for proton transport of about 2 kcal/mol over a large range of humidity conditions. The diffusion being in the same range as proton transport in Nafion, embedded functionalised silicon dioxide particles are expected to actively participate in the proton transport and provide further pathways through the membrane.

Conclusion and Discussion

The progress in computational techniques and computer resources achieved during the past decades enables computational research to give significant contributions in the field of applied material design. The gain of interdisciplinarity is a deeper understanding of the interplay between material properties and atomistic mechanisms, a complex issue that demands a careful interpretation of experimental measurements and model building, as well as computer simulation on different length scales.

The present study describes a class of additives to a polymer electrolyte membrane for fuel cell application. The material consists of organically functionalised MCM-41 silicon dioxide, the functional groups being sulphonic acid or imidazole. Water retention and enhancement of proton conductivity is credited to the material. Both these properties were approached by computational methods.

The water adsorption of a material is usually characterised by measurements of isotherms, but to evaluate the effect of functionalisation on the atomistic scale, a slab model of silicon dioxide with different surface coverage was studied by classical molecular dynamic simulations (Sec. 4.1.2). The choice of a simple slab model described by FF allows to span large time and length scales, which are important for this type of parameter. The simulations of slab models provide an estimate for the local chemical environment inside the porous material under different humidity conditions. It was shown that sulphonic acid enhances the hydrophilic character of the silicon dioxide material and that the mobility and the aggregation of both functional groups are significantly influenced by the chemical environment.

An accurate analysis of proton transport requires a correct description of chemical reactions and charge transfer processes, which is accomplished by quantum mechanical methods. To avoid the use of very time consuming methods, a semi-empirical approach to the electronic Hamiltonian was chosen. The DFTB approach showed a sufficient accuracy for the energetics of proton conducting groups (Sec. 3.1) and is sufficiently fast to provide an adequate sampling of the systems. Nevertheless, systems described by DFTB are reduced to a minimal size to save more computer time and, if needed, the chemical environment is included in the simulation through the FF description using a QM/MM approach.

From preliminary studies of functional groups in vacuum (Sec. 3.2.2), the direct proton transport between the anchored functional groups is known. The mechanism was identified to consist of two steps, proton hopping or structural proton diffusion between aggregated proton conducting groups, followed by a local vehicular proton transport depending on the cleavage of hydrogen bonds and formation of new aggregates. This mechanism highly depends on the density of groups and is significantly suppressed under the influence of the chemical environment inside the pore (Sec. 5.3).

Beside the direct proton transport, two additional mechanisms were considered in the porous environment. In the limit of the full hydrated system a purely water based proton transport is expected, while in low humidity conditions an intermediate mechanism of water based transport between the proton conducting groups is considered. The three mechanisms were evaluated for the sulphonic acid and the imidazole system by calculation of free energy barriers.

To estimate the occurrence of the purely water based transport, deprotonation of proton conducting groups was studied. In agreement with experimental results, it was found that the protonated state of imidazole is favoured by about 12 kcal/mol even at high humidity. Deprotonation of sulphonic acid is instead expected at low hydration. The subsequent water based proton transport shows similar diffusion coefficient as hydrated Nafion.

In low hydration conditions, transport between sulphonic acid groups via several water molecules is enabled and dominates over the direct proton transport, as hydrogen bond fluctuations are low. The proton transport barrier between sulphonic acid groups strongly decreases with increasing water content, whereas in the imidazole system a weaker dependence on the water content is expected. For imidazole, the barrier for the direct proton transport mechanism is lower by a factor of two than for the water based transport.

The aim of the project was to improve and understand the proton transport of the PEM material in low water conditions and at elevated temperatures. In polymer material, e.g. Nafion, the structure of the water channels highly depends on the humidity. In contrast to this, the porous structure of MCM-41 is thermodynamically stable and computer simulation showed a complete wetting of the pore surface even under low humidity conditions that enable a low barrier proton transport between sulphonic acid molecules, while at higher water content the sulphonic acid groups are deprotonated and transport is water based. By conductivity measurements (Sec. 2.2.2), a strong effect of water content and temperature in sulphonic acid based systems was observed that could hint to a cross-over between different mechanisms as suggested by computer simulation. The performance of imidazole based systems is less dependent on the water content and therefore, accomplishes to some extent even better the objective. On the other hand, low intrinsic charge carrier concentration leads to low conductivities and the diffusion coefficients estimated in a system of functional groups in vacuum were at least one order of magnitude lower than diffusion in bulk water. Especially in the imidazole system, proton transport abilities at dry conditions strongly depend on a homogeneous functionalisation and a high density of functional groups[i].

---

[i]Such material has not been synthesised yet. So far only grafted functionalised imidazole system were measured with a relatively low surface density and inhomogeneous distribution of functional molecules.

Appendix A
Additional Data

The following data belongs to Sec. 3.1.

Table A.1: *Binding energy of water cluster from different level of theory: post Hartree-Fock method MP2, DFT with B3LYP functional, DFTB method with SK-file set pbc/mio with and without hydrogen bond description. DFT and MP2 using different Basis function: cc-pV(D/T/Q)Z and 6-31g*. (◇ data taken from [111])*

method	basis set	MP2	DFT	DFTB	H-bond-DFTB
water(2)	cc-pVDZ	3.8	4.1	1.7	2.3
water(2)	cc-pVTZ	3.1	2.9		
water(2)	cc-pVQZ		2.6		
water(2)	6-31g*		3.8		
water(2)	6-31g(d,p)		3.8 ◇		
water(3)	cc-pVDZ	8.1	9.0	3.3	4.8
water(3)	cc-pVTZ	6.5	6.2		
water(3)	cc-pVQZ		5.4		
water(3)	6-31g*		8.5		
water(3)	6-31g(d,p)		8.3 ◇		
water(4)	cc-pVDZ		11.1	4.4	6.2
water(4)	cc-pVTZ		8.1		
water(4)	cc-pVQZ		7.2		
water(4)	6-31g*		10.6		
water(4)	6-31g(d,p)		10.4 ◇		
water(10)	cc-pVDZ	13.8	15.0	5.9	8.5
water(10)	cc-pVTZ		10.8		
water(10)	cc-pVQZ		9.1		
water(10)	6-31g(d,p)		14.0 ◇		
water(15)	cc-pVDZ		16.0	6.4	9.3
water(15)	cc-pVTZ		11.4		
water(15)	6-31g*		15.1		
water(15)	6-31g(d,p)		14.9 ◇		
water(20)	cc-pVDZ		16.8	6.7	9.8

Table A.2: *Proton affinities of water cluster, methyl imidazole (MI) and methyl sulphonic acid (MSA) from different level of theory: post Hartree-Fock method MP2, DFT with B3LYP functional, DFTB method with SK-file set pbc/mio with and without hydrogen bond description. DFT and MP2 using different Basis function: cc-pV(D/T/Q)Z and 6-31g*.*

method	basis set	MP2	DFT	DFTB	H-bond-DFTB
MI	cc-pVDZ	239.6	241.9	234	235
MI	cc-pVTZ		240.9		
MI	cc-pVQZ		240.3		
MSA				187	187
MSA(-)	cc-pVDZ	341.8	340.1	333	335
MSA(-)	cc-pVTZ		340.6		
MSA(-)	6-31g*		333.9		
water(1)	cc-pVDZ	179.7	178.3	180	177
water(1)	cc-pVTZ	174.3	173.9		
water(1)	cc-pVQZ		172.3		
water(1)	6-31g*		175.8		
water(2)	cc-pVDZ	211.5	210.6	206	207
water(2)	cc-pVTZ	205.6	204.4		
water(2)	cc-pVQZ		202.1		
water(2)	6-31g*		204.9		
water(3)	cc-pVDZ	222.5	221.6	217	217
water(3)	cc-pVTZ	217.6	218.0		
water(3)	cc-pVQZ		216.2		
water(3)	6-31g*		219.0		
water(4)	cc-pVDZ		228.5	226	225
water(4)	cc-pVTZ		225.1		
water(4)	cc-pVQZ		223.5		
water(4)	6-31g*		225.7		
water(10)	cc-pVDZ	244.4	242.0	239	240
water(10)	cc-pVTZ		239.3		
water(10)	cc-pVQZ		237.8		
water(10)	6-31g*		241.5		
water(15)	cc-pVDZ		246.1	244	243
water(15)	cc-pVTZ		243.6		
water(15)	6-31g*		244.3		
water(20)	cc-pVDZ			246	244

Appendix B

Changes in Gromacs 4.0.5

In the course of this work, the Gaussian-QM/MM implementation in Gromacs 4.0.5. was adopted for coupling to the dftb+ code for the QM part and the mCEC coordinate was implemented in the program for the use of umbrella sampling for proton transport. This appendix will document shortly usage and changes compared to the Gromacs 4.0.5 official version

B.1 Usage of the dftb+ QM/MM

The usage of the modified QM/MM coupling in Gromacs is explained in addition to instructions in Gromacs manual to create the correct QM/MM-topology and gro-file containing link atoms and constraints if needed. As the QM/MM implementation in Gromacs 4.0.5 only works with periodic boundary conditions, the same holds for the modified version.

mdp-input:

The following keywords must be added to the molecular-dynamics-parameter input file, **mdp-input**:

```
QMMM=yes
QMMM-grps= QM_name
MMChargeScaleFactor=1.0
QM_charge = 0.0
```

As in Gromacs 4.0.5, the keywords (*QMMM*) specifies the request of a QM/MM simulation and the QM atoms are listed as a group (*QM_name*) by their indices in the so called index file (**ndx**). The scale factor (*MMChargeScaleFactor*) can be specified to scale the electrostatic interaction between the MM point charges and the QM atoms in the QM calculation.

While starting the simulation, the environment variable **DFTB_EXE** should point to the *dftb+_1.0* binary. The dftb+ input file is then read in automatically. If there is no such input file, called *dftb_in.hsd*, a default dftb+ input file is generated. In this case, the path to the SK-file directory should be specified in the environment variable **DFTB_SK** and the charge of the QM-zone should be set correctly.

During the mdrun, Gromacs writes the coordinate file of the QM zone*dftb_input_geometry.gen* in the gen-fileformat[42][115] and the file *dftb_external_charges*, which contains the coordinates and the charges (multiplied by the scale factor) of the MM atoms. Then it calls the path specified in the environment variable DFTB_EXE. The DFTB output is written in the file *dftb.out*.

input - dftb+:

As mentioned above, the dftb+ input file *dftb_in.hsd* is generated automatically or should be placed in the directory where the calculation takes place. It containes the following lines to correctly read in the files generated by Gromacs:

```
Geometry = GenFormat {
<<< dftb_input_geometry.gen
}
```

and

```
  ElectricField = {
    PointCharges = {
      CoordsAndCharges [Angstrom] = {
<<< dftb_external_charges
      }
    }
  }
```

The following command enables the calculation of forces that are written in *results.tag* by *dftb+_1.0* and read in by Gromacs afterwards:

`CalculateForces=yes`

The details of the actual *dftb+_1.0* input as well as the Gromacs input is explained in detail in the manuals of both programs.

B.2   Usage of the mCEC-coordinate

The mCEC coordinate is designed to denote the proton position in a MD simulation. The coordinate was implemented earlier in CHARMM program. Further details about the coordinate are given in Sec. 1.4. To implement umbrella sampling with the mCEC coordinate as a reference, the *pull* code of Gromacs 4.0.5 was modified. To use the changed *pull*-code, the following keywords must be added to the molecular-dynamics-parameter input file.

mdp-input:

```
pull = umbrella_mCEC
pull_k_theta0=300
pull_r_Null0=0.13
pull_group0= GROUP_NAME
pull_weights0= 1.0 1.0 1.0 ...
nr_of_mCEC_term=2
```

The keyword *umbrella_mCEC* enables the use of the modified implementation, all other pull options are not influenced by the new implementation.

The parameters *pull_k_theta1*= k^1 and *pull_r_Null1*= k^0 define the switching function as explained in Sec. 1.4. All the groups (*pull_group0* or *pull_group1* ...) are mCEC group with weight w^A listed in *pull_weights0* (or *pull_weights1*), each atom in the group must be assigned a weight. For hydrogen atoms the weight is set to 1.0, otherwise the correction terms (*nr_of_mCEC_term*$\geq$0) are not calculated in the correct way. The parameter *nr_of_mCEC_term*=N specifies which terms of the mCEC coordinate should be used. 0 - just the first, 1 - the first two terms plus a correcetion to the first term due to periodic boundary conditions and 2 - all terms plus another correction due to pbc. For the reference group usually only the first term is needed!

Another option has to be specified. The *pull_geometry* can be set to *distance* or *ratio*. The reaction coordinate for the request *distance* is the distance between only two groups, while in the case of *ratio* one has to specify 3 groups. Here, the reaction coordinate is the relative distance between the mCEC coordinate (defined by group0) and the two reference groups, defined by group1 and group2. For details see Sec. 5.1.

output:

The **output** of the calculation is a file called *mCEC*. It contains time, actual value of the reaction coordinate and the reference value of the reaction coordinate. In addition, the files *pullf.xvg* and *pullx.xvg* - known from Gromacs manual - are created.

additional tool:

The **gmx_mcec tool** reads in xtc-file (-f) and tpr-file (-s), with a correctly defined it umbrella_mCEC-pull part. This tool will create two files, the coordinate file *mcec.gro* and the trajectory file *mcec.xtc*, which contain the trajectory of each Pull-group defined in the tpr-file.

B.3 Details of the Implementation

B.3.1 Derivatives of the mCEC

The implementation of a proton coordinate (mCEC=$\vec{c}$; see Sec. 1.4) in a MD simulation program enables to use the proton coordinate as part of a reaction coordinate. To distribute the forces over the atoms that are involved in the proton coordinate, the following derivatives are needed. The derivative of the mCEC coordinate $\partial_l c_m$ is distinguished between the derivative according to the coordinates of a hydrogen atom H $\partial_l^H c_m$ (see Eq. B.2) and the derivative according to the coordinates of a non-hydrogen atom X $\partial_l^X c_m$ (see Eq. B.1).

$$\begin{aligned}
\partial_l^X c_m &= w^X \delta_{ml} + \vartheta^{XH} \delta_{ml} + (R_m^X - R_m^H) \partial_l^X \vartheta^{XH} \\
&+ (w^X \vartheta_{max}^{XH'} - w^{X''} \vartheta_{max}^{X''H''}) \delta_{ml} + w^X (R_m^X - R_m^{X''}) \partial_l^X \vartheta_{max}^{XH'} \\
&\forall \ X, X'' \in \text{(same molecule)} \\
\partial_l^H c_m &= w^X \delta_{ml} - \vartheta^{XH} \delta_{ml} + (R_m^X - R_m^H) \partial_l^H \vartheta^{XH} \\
&+ w^{X'} (R_m^{X'} - R_m^{X''}) \partial_l^H \vartheta_{max}^{X'H'} \delta^{HH'} \\
&+ w^{X''} (R_m^{X''} - R_m^{X'}) \partial_l^H \vartheta_{max}^{X''H''} \delta^{HH''}
\end{aligned}$$

(B.1)

(B.2)

These expressions depend on the partial derivatives of the switching function according to non-hydrogen (X) coordinates (see Eq. B.3) and to the hydrogen (H) (see Eq. B.4):

$$\partial_l^X \vartheta^{XH} = -k^1 \frac{\exp(k^1 |R^X - R^H| - k^0 k^1)}{(1 + \exp(k^1 |R^X - R^H| - k^1 k^0))^2} \cdot \frac{(R^X - R^H)_l}{|R^X - R^H|} \quad (B.3)$$

$$\partial_l^H \vartheta^{XH} = k^1 \frac{\exp(k^1 |R^X - R^H| - k^0 k^1)}{(1 + \exp(k^1 |R^X - R^H| - k^1 k^0))^2} \cdot \frac{(R^X - R^H)_l}{|R^X - R^H|} \quad (B.4)$$

B.3.2 Derivatives of the Reaction Coordinates

The following formulas and derivatives are needed for the implementation, in addition to Sec. 5.1.

The umbrella sampling potential (P) is applied to the *distance reaction coordinate* as described below and the forces ($F_l^X = -\partial_l^X P$) are explicitly calculated:

$$\begin{aligned}
P &= 0.5 k^u (\zeta^d - I)^2 \\
F_l^X &= -\partial_l^X P \\
&= -k^u (\zeta^d - I) \partial_l^X \zeta^d \\
&= -k^u (\zeta^d - I) \frac{(c_m - R_m^0)}{\zeta^d} (\partial_l^X c_m - \partial_l^X R_m^0)
\end{aligned}$$

The umbrella sampling potential (P) is applied to the *ratio reaction coordinate* as described below and the forces ($F_l^X = -\partial_l^X P$) are explicitly calculated:

$$\begin{aligned}
P &= 0.5 k^u (\zeta^r - I)^2 \\
F_l^X &= -\partial_l^X P \\
&= -k^u (\zeta^r - I) \partial_l^X \zeta^r \\
\partial_l^X \zeta^r &= \sum_m \frac{(c_m - R_m^A)}{\zeta_A^d} (\zeta_A^d + \zeta_B^d)^{-2} (2\zeta_B^d)(\partial_l^X c_m - \partial_l^X R_m^A) \\
&+ \sum_m \frac{(c_m - R_m^B)}{\zeta_B^d} (\zeta_A^d + \zeta_B^d)^{-2} (-2\zeta_A^d)(\partial_l^X c_m - \partial_l^X R_m^B)
\end{aligned}$$

Appendix C
Force Field Parameters

All FF parameters as introduced in Sec. 1.1.1 are listed in the following. The FF parameters were used for the classical MD simulations described in Sec. 3.2.2 and Sec. 4.1.2.

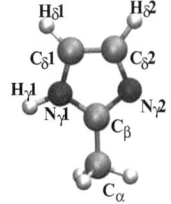

typ		$r_0^{bond,(AB)}$	$k_{(AB)}^{bond}$
A	B	[nm]	[kJ/mol]
H$_2$O (TIP3P) and H$_3$O$^+$			
O	H	0.09572	502416.0
silicon dioxide			
Si	O	0.161	251040.0
O	H	0.09572	502080
Si	C	0.186	224262.4
carbon chain			
C	C	0.151	224262.4
C	H	0.109	284512.000
neutral/protonated imidazole			
C$_\beta$	N$_{\delta 1}$	0.1343	399153.6
C$_\beta$	N$_{\delta 2}$	0.1335	408358.4
N$_{\gamma 1}$	H$_{\gamma 1}$	0.101	363171.2
N$_\gamma$	C$_\delta$	0.13940	343088.0
C$_\delta$	C$_\delta$	0.137	435136.0
C$_\delta$	H$_\delta$	0.10900	284512.0
neutral/deprotonated sulphonic acid			
S	C	0.181	185769.6
S	O	0.144	585760.0
O	H	0.0945	462750.4

Table C.1: *FF parameters: Bonds, see Eq. 1.2. The atom nomenclature for the imidazole molecule is indicated beside.*

	typ		$\theta_0^{angle,(ABC)}$	$k_{(ABC)}^{angle}$
C	B	C	[degree]	[kJ mol^{-1} rad^{-2}]
H$_2$O (TIP3P) and H$_3$O$^+$				
H	O	H	113.52	628.02
silicon dioxide				
Si	O	Si	148.	397.480
O	Si	O	109.	397.480
carbon chain				
C	C	C	112.7	488.273
C	C	H	110.7	313.800
H	C	H	107.8	276.144
neutral/protonated imidazole				
C	C$_\alpha$	C$_\beta$	114.	527.184
N$_{\gamma 2}$	C$_\beta$	N$_{\gamma 1}$	120.000	585.760
C$_\delta$	N$_{\gamma 1}$	C$_\beta$	109.800	585.760
C$_\beta$	N$_{\gamma 2}$	C$_{\delta 2}$	110.000	585.760
C$_{\delta 2}$	C$_{\delta 1}$	N$_{\gamma 1}$	106.300	585.760
C$_{\delta 1}$	C$_{\delta 2}$	N$_{\gamma 2}$	111.000	585.760
C$_\alpha$	C$_\beta$	N$_\gamma$	125.000	585.760
C$_{\delta 2}$	C$_{\delta 1}$	H$_{\delta 1}$	130.700	292.880
C$_{\delta 1}$	C$_{\delta 2}$	H$_{\delta 2}$	128.200	292.880
N$_\gamma$	C$_\delta$	H$_\delta$	121.600	292.880
C$_{\delta 1}$	N$_{\gamma 1}$	H$_{\gamma 1}$	120.000	292.880
neutral/deprotonated sulphonic acid				
O	S	C	108.900	619.232
H	C	S	109.500	292.880
O	S	O	119.00	870.272
S	O	H	119.00	870.272

Table C.2: *FF parameters: Angles, see Eq. 1.3.*

	typ			$k_0^{(ABC)}$	$k_1^{(ABC)}$	$k_2^{(ABC)}$	$k_3^{(ABC)}$	$k_4^{(ABC)}$	$k_5^{(ABC)}$
A	B	C	D			[kJ/mol]			
				carbon chain					
H	C	C	H	0.6276	1.8828	0.0000	-2.5104	0.0	0.0
C	C	C	H	0.6276	1.8828	0.0000	-2.5104	0.0	0.0
C	C	C	C	2.9288	-1.4644	0.2092	-1.6736	0.0	0.0
				neutral/protonated imidazole					
X	$C_{\delta 1}$	$C_{\delta 2}$	X	44.97800	0.0	-44.9780	0.0	0.0	0.0
X	$C_{\delta 1}$	$N_{\gamma 1}$	X	11.71520	0.0	-11.7152	0.0	0.0	0.0
X	C_α	$N_{\gamma 2}$	$C_{\delta 2}$	41.84000	0.0	-41.8400	0.0	0.0	0.0
X	C_α	$N_{\gamma 1}$	$C_{\delta 1}$	19.45560	0.0	-19.4556	0.0	0.0	0.0
C_δ	N_γ	C_α	N_γ	19.45560	0.0	-19.4556	0.0	0.0	0.0
$H_{\gamma 1}$	$N_{\gamma 1}$	C_α	X	13.38880	0.0	-13.3888	0.0	0.0	0.0
$H_{\delta 2}$	$C_{\delta 2}$	$N_{\gamma 2}$	C_α	13.38880	0.0	-13.3888	0.0	0.0	0.0
$H_{\gamma 1}$	$N_{\gamma 1}$	C_α	X	13.38880	0.0	-13.3888	0.0	0.0	0.0
X	C_δ	N_γ	X	20.08320	0.0	-20.0832	0.0	0.0	0.0
C_δ	C_δ	N_γ	C_α	-3.55640	2.092	1.4644	0.0	0.0	0.0
				neutral/deprotonated sulphonic acid					
H	C	C	S	0.7322	2.1966	0.0	-2.9288	0.0	0.0
C	C	S	O	0.0	0.0	0.0	0.0	0.0	0.0
H	C	S	O	0.7322	2.1966	0.0	-2.9288	0.0	0.0

Table C.3: *FF parameters: Dihedrals, see Eq. 1.5.*

atom type A	number of atoms per molecule	q_A [e]	m_A [a.m.u.]	σ_A [kJ/mol]	ϵ_A [nm]
H_3O^+					
O	1	-0.620	9.95140	0.315061	0.636386
H	3	0.540	4.03200	0.00000	0.00000
H_2O (tip3p)					
O	1	-1.080	9.95140	0.315061	0.636386
H	2	0.540	4.03200	0.0000	0.000
silicon dioxide					
Si	≈1	1.08000000	28.08000	0.392	2.51
O	≈2	-0.5586	15.99940	0.3154	0.6368
H		0.32000000	1.00800	0.04	0.1926
carbon chain CH_2 - group					
C	1	0.12	12.01100	0.35000	0.276144
H	2	0.06	1.00800	0.25000	0.125520
carbon chain CH_3- group					
C	1	0.18	12.01100	0.35000	0.276144
H	3	0.06	1.00800	0.25000	0.125520
imidazole					
C_β	1	0.49	12.011	0.355	0.29288
$N_{\gamma 1}$	1	-0.30	14.00670	0.325	0.711280
$N_{\gamma 2}$	1	-0.6	14.00670	0.32500	0.711280
$C_{\delta 2}$	1	0.10	12.01100	0.3550	0.29288
$C_{\delta 1}$	1	-0.27	12.01100	0.3550	0.29288
$H_{\gamma 1}$	1	0.31	1.00800	0.00000	0.00000
$H_{\delta 1}$	1	0.16	1.00800	0.2420	0.12552
$H_{\delta 2}$	1	0.11	1.00800	0.2420	0.12552
protonated imidazole					
C_β	1	0.7	12.011	0.355	0.29288
$N_{\gamma 1}$	2	-0.38	14.00670	0.325	0.71128
$C_{\delta 1}$	2	-0.1	12.01100	0.355	0.29288
$H_{\gamma 1}$	2	0.40	1.00800	0.2420	0.12552
$H_{\delta 1}$	2	0.23	1.00800	0.0	0.0
neutral sulphonic acid					
S	1	1.65	36.06000	0.355	1.0465
O	2	-0.69	15.99940	0.3154	0.8372
O	1	-0.73	15.99940	0.3154	0.86368
H	1	0.46	1.00800	0.00	0.00
deprotonated sulphonic acid					
S	1	1.25	36.06000	0.355	1.0465
O	3	-0.75	15.99940	0.3154	0.8372

Table C.4: *FF parameters: Atom types, see Eq. 1.1.*

Bibliography

[1] P. Tölle and C. Köhler. Water free proton transport in imidazole functionalised silicon dioxide material - calculation of free energy barrier dependent on the mcec proton coordinate. *submitted to physica status solidi*, 2011.

[2] L. Gubler and G.G. Scherer. Trends for fuel cell membrane development. *Desalination*, 250(3):1034–1037, 2010.

[3] J. Zhang, Z. Xie, J. Zhang, Y. Tang, C. Song, T. Navessin, Z. Shi, D. Song, H. Wang, D.P. Wilkinson, Z.-S. Liu, and S. Holdcroft. High temperature PEM fuel cells. *Journal of Power Sources*, 160(2):872–891, 2006.

[4] W. Goddard, B. Merinov, A. Van Duin, T. Jacob, M. Blanco, V. Molinero, SS Jang, and YH Jang. Multi-paradigm multi-scale simulations for fuel cell catalysts and membranes. *Molecular Simulation*, 32(3):251–268, 2006.

[5] K.D. Kreuer, S.J. Paddison, E. Spohr, and M. Schuster. Transport in proton conductors for fuel-cell applications: Simulations, elementary reactions, and phenomenology. *Chemical Reviews*, 104(10):4637–4678, 2004.

[6] M.F.H. Schuster and W.H. Meyer. Anhydrous Proton-Conducting Polymers. *Annual Review of Materials Research*, 33(1):233–261, 2003.

[7] D.R. Morris and X. Sun. Water-sorption and transport properties of Nafion 117 H. *Journal of Applied Polymer Science*, 50(8):1445–1452, 1993.

[8] T.D. Gierke and W.Y. Hsu. The cluster-network model of ion clustering in perfluorosulfonated membranes. 180:283–307, 1982.

[9] K.A. Mauritz and R.B. Moore. State of understanding of Nafion. *Chemical Review*, 104(10):4535–4586, 2004.

[10] N.P. Blake, M.K. Petersen, G.A. Voth, and H. Metiu. Structure of hydrated Na- Nafion polymer membranes. *Journal of Physical Chemistry B*, 109(51):24244–24253, 2005.

[11] M.A. Hickner, H. Ghassemi, Y.S. Kim, B.R. Einsla, and J.E. McGrath. Alternative polymer systems for proton exchange membranes (PEMs). *Chemical Reviews*, 104(10):4587–4612, 2004.

[12] W.L. Jorgensen, D.S. Maxwell, and J. Tirado-Rives. Development and testing of the OPLS all-atom force field on conformational energetics and properties of organic liquids. *Journal of the American Chemical Society*, 118(45):11225–11236, 1996.

[13] P.P. Ewald. Die Berechnung optischer und elektrostatischer Gitterpotentiale. *Annalen der Physik*, 369(3):253–287, 1921.

[14] M.P. Allen and D.J. Tildesley. *Computer simulation of liquids*. Oxford University Press, 2009 (1987).

[15] S.W. De Leeuw, J.W. Perram, and E.R. Smith. Simulation of electrostatic systems in periodic boundary conditions. I. Lattice sums and dielectric constants. *Proceedings of The Royal Society of London. Series A*, 373(1752):27–56, 1980.

[16] T. Darden, D. York, and L. Pedersen. Particle Mesh Ewald: An N log (N) method for Ewald sums in large systems. *The Journal of Chemical Physics*, 98:10089, 1993.

[17] V. Fock. Näherungsmethode zur Lösung des quantenmechanischen Mehrkörperproblems. *Zeitschrift für Physik A*, 61(1):126–148, 1930.

[18] J.C. Slater. A simplification of the Hartree-Fock method. *Physical Review*, 81(3):385–390, 1951.

[19] S.M. Blinder. Basic concepts of self-consistent-field theory. *American Journal of Physics*, 33(6):431, 1965.

[20] A. Szabo and N.S. Ostlund. *Modern quantum chemistry*. Dover Publications, New York, 1989.

[21] J. C. Slater. Wave functions in a periodic potential. *Physical Review*, 51(10):846–851, 1937.

[22] C. Møller and M.S. Plesset. Note on an approximation treatment for many-electron systems. *Physical Review*, 46(7):618–622, 1934.

[23] P. Hohenberg and W. Kohn. Inhomogeneous electron gas. *Physisical Review B*, 136(3):864–871, 1964.

[24] W. Kohn, L.J. Sham, et al. Self-consistent equations including exchange and correlation effects. *Physical Review A*, 140(4):1133–1138, 1965.

[25] G. Czycholl. *Theoretische Festkörperphysik*. Springer-Verlag, 2008.

[26] A.D. Becke. A new mixing of Hartree–Fock and local density-functional theories. *The Journal of Chemical Physics*, 98(2):1372, 1993.

[27] D. Porezag, T. Frauenheim, T. Köhler, G. Seifert, and R. Kaschner. Construction of tight-binding-like potentials on the basis of density-functional theory: Application to carbon. *Physical Review B*, 51(19):12947–12957, 1995.

[28] T. Frauenheim, G. Seifert, M. Elstner, Z. Hajnal, G. Jungnickel, D. Porezag, S. Suhai, and R. Scholz. A self-consistent charge density-functional based tight-binding method for predictive materials simulations in physics, chemistry and biology. *Computer simulation of materials at atomic level*, 217(1):41, 2000.

[29] W. Matthew C. Foulkes and Roger Haydock. Tight-binding models and density-functional theory. *Physical Review B*, 39(17):12520–12536, 1989.

[30] I.N. Bronstein, KA Semendjajew, G. Musiol, and H. Mühlig. *Taschenbuch der Mathematik*. B.-G. Teubner Stuttgart - Leipzig, 1991.

[31] J.C. Slater. Atomic shielding constants. *Physical Review*, 36(1):57–64, 1930.

[32] D. Porezag, Th. Frauenheim, Th. Köhler, G. Seifert, and R. Kaschner. Construction of tight-binding-like potentials on the basis of density-functional theory: Application to carbon. *Physical Review B*, 51(19):12947–12957, 1995.

[33] M. Elstner, D. Porezag, G. Jungnickel, J. Elsner, M. Haugk, T. Frauenheim, S. Suhai, and G. Seifert. Self-consistent-charge density-functional tight-binding method for simulations of complex materials properties. *Physical Review B*, 58(11):7260–7268, 1998.

[34] H. Hu, Z. Lu, M. Elstner, J. Hermans, and W. Yang. Simulating water with the SCCDFTB method: from molecular clusters to the liquid state. *The journal of physical chemistry. A*, 111(26):5685, 2007.

[35] M. Elstner. The SCC-DFTB method and its application to biological systems. *Theoretica Chimica Acta*, 116(1):316–325.

[36] A. Warshel and M. Levitt. Theoretical studies of enzymic reactions: dielectric, electrostatic and steric stabilization of the carbonium ion in the reaction of lysozyme. *Journal of molecular biology*, 103(2):227–249, 1976.

[37] U.C. Singh and P.A. Kollman. A combined ab initio quantum mechanical and molecular mechanical method for carrying out simulations on complex molecular systems: Applications to the CH3Cl+ Cl-exchange reaction and gas phase protonation of polyethers. *Journal of Computational Chemistry*, 7(6):718–730, 1986.

[38] M.J. Field, P.A. Bash, and M. Karplus. A combined quantum mechanical and molecular mechanical potential for molecular dynamics simulations. *Journal of Computational Chemistry*, 11(6):700–733, 1990.

[39] H.M. Senn and W. Thiel. QM/MM methods for biomolecular systems. *Angewandte Chemie International Edition*, 48(7):1198–1229, 2009.

[40] JM Knaup, P. Tölle, C. Köhler, and T. Frauenheim. Quantum mechanical and molecular mechanical simulation approaches bridging length and time scales for simulation of interface reactions in realistic environments. *The European Physical Journal*, 177(1):59–81, 2009.

[41] D. Van Der Spoel, E. Lindahl, B. Hess, G. Groenhof, A.E. Mark, and H.J.C. Berendsen. GROMACS: fast, flexible, and free. *Journal of computational chemistry*, 26(16):1701–1718, 2005.

[42] Density Functional Tight Binding website. *www.dftb-plus.info*.

[43] L. Verlet. Computer experiments on classical fluids - II. Equilibrium correlation functions. *Phys. Rev*, 165(1):201–14, 1968.

[44] P.S.Y. Cheung. On the calculation of specific heats, thermal pressure coefficients and compressibilities in molecular dynamics simulations. *Molecular Physics*, 33(2):519–526, 1977.

[45] Eugene N. Parker. Tensor virial equations. *Physical Review*, 96(6):1686–1689, 1954.

[46] W.G. Hoover. Canonical dynamics: Equilibrium phase-space distributions. *Physical Review A*, 31(3):1695–1697, 1985.

[47] H.C. Andersen. Molecular dynamics simulations at constant pressure and/or temperature. *J. Chem. Phys*, 72(4):2384–2393, 1980.

[48] M. Parrinello and A. Rahman. Polymorphic transitions in single crystals: A new molecular dynamics method. *Journal of Applied Physics*, 52(12):7182, 1981.

[49] H.J.C. Berendsen, J.P.M. Postma, W.F. Van Gunsteren, A. DiNola, and J.R. Haak. Molecular dynamics with coupling to an external bath. *The Journal of Chemical Physics*, 81(8):3684, 1984.

[50] B. Roux. The calculation of the potential of mean force using computer simulations. *Computer Physics Communications*, 91(1-3):275–282, 1995.

[51] S. Kumar, J.M. Rosenberg, D. Bouzida, R.H. Swendsen, and P.A. Kollman. The weighted histogram analysis method for free-energy calculations on biomolecules. I. The method. *Journal of Computational Chemistry*, 13(8), 1992.

[52] C. Chipot and A. Pohorille. *Free energy calculations: theory and applications in chemistry and biology*. Springer Verlag, 2007.

[53] P.H. Konig, N. Ghosh, M. Hoffmann, M. Elstner, E. Tajkhorshid, T. Frauenheim, and Q. Cui. Toward Theoretical Analysis of Long-Range Proton Transfer Kinetics in Biomolecular Pumps. *Journal of Physical Chemistry A*, 110(2):548–563, 2006.

[54] P. Tölle, W.L. Cavalcanti, M. Hoffmann, C. Köhler, and T. Frauenheim. Modelling of Proton Diffusion in Immobilised Imidazole Systems for Application in Fuel Cells. *Fuel Cells*, 8(3):236–243, 2008.

[55] E. Wicke, M. Eigen, and T. Ackermann. Über den Zustand des Protons (Hydroniumions) in wäßriger Lösung. *Zeitschrift für Physikalische Chemie*, 1:343–364, 1954.

[56] M. Eigen and E. Wicke. The thermodynamics of electrolytes at higher concentration. *The Journal of Physical Chemistry*, 58(9):702–714, 1954.

[57] G. Zundel and H. Metzger. Energiebänder der tunnelnden Uberschuss-Protonen in flüssigen Säuren. Eine IR-spektroskopische Untersuchung der Natur der Gruppierungen H5O2+. *Zeitschrift für Physikalische Chemie*, 58:225, 1968.

[58] C.J.T. De Grotthuss. Sur la décomposition de l'eau et des corps qu'elle tient en dissolution à l'aide de l'électricité galvanique. *Ann. Chim*, 58:54–74, 1806.

[59] L.I. Yeh, M. Okumura, J.D. Myers, J.M. Price, and Y.T. Lee. Vibrational spectroscopy of the hydrated hydronium cluster ions $H_3O^+\cdot (H_2O)_n$, (n= 1, 2, 3). *J. Chem. Phys*, 91(12).

[60] U.W. Schmitt and G.A. Voth. The computer simulation of proton transport in water. *The Journal of Chemical Physics*, 111(20):9361, 1999.

[61] D. Marx, M.E. Tuckerman, J. Hutter, and M. Parrinello. The nature of the hydrated excess proton in water. *Nature*, 397(6720):601–604, 1999.

[62] M.E. Tuckerman, D. Marx, and M. Parrinello. The nature and transport mechanism of hydrated hydroxide ions in aqueous solution. *Nature*, 417(6892):925–929, 2002.

[63] K. Kreuer. Fast proton transport in solids. *Journal of Molecular Structure*, 177:265–276, 1988.

[64] K.D. Kreuer. Fast proton conductivity: A phenomenon between the solid and the liquid state? *Solid State Ionics*, 94(1-4):55–62, 1997.

[65] M.E. Tuckerman, D. Marx, M.L. Klein, and M. Parrinello. On the quantum nature of the shared proton in hydrogen bonds. *Science*, 275(5301):817, 1997.

[66] RA Robinson and RH Stokes. *Electrolyte solutions,*. Butterworths, London, 1959.

[67] W.Q. Deng, V. Molinero, and W.A. Goddard III. Fluorinated imidazoles as proton carriers for water-free fuel cell membranes. *Journal of the American Chemical Society*, 126(48):15644–15645, 2004.

[68] S. Izvekov and G.A. Voth. Ab initio molecular-dynamics simulation of aqueous proton solvation and transport revisited. *The Journal of chemical physics*, 123:44505, 2005.

[69] T.J.F. Day, A.V. Soudackov, M. Čuma, U.W. Schmitt, and G.A. Voth. A second generation multistate empirical valence bond model for proton transport in aqueous systems. *The Journal of Chemical Physics*, 117(12):5839, 2002.

[70] M.F.H. Schuster and W.H. Meyer. Anhydrous proton-conducting polymers. *Materials Research*, 33(1):233, 2003.

[71] J.T. Daycock, G.P. Jones, J.R.N. Evans, and J.M. Thomas. Rotation of Imidazole in the Solid State and its Significance in deciding the Nature of Charge Migration in Biological Materials. *Nature*, 218:672–673, 1968.

[72] B.S. Hickman, M. Mascal, J.J. Titman, and I.G. Wood. Protonic conduction in imidazole: a solid-state 15N NMR study. *Journal of the American Chemical Society*, 121(49):11486–11490, 1999.

[73] W. Münch, K.D. Kreuer, W. Silvestri, J. Maier, and G. Seifert. The diffusion mechanism of an excess proton in imidazole molecule chains: first results of an ab initio molecular dynamics study. *Solid State Ionics*, 145(1-4):437–443, 2001.

[74] K.D. Kreuer, A. Fuchs, M. Ise, M. Spaeth, and J. Maier. Imidazole and pyrazole-based proton conducting polymers and liquids. *Electrochimica Acta*, 43(10-11):1281–1288, 1998.

[75] A. Kawada, A.R. McGhie, and M.M. Labes. Protonic conductivity in imidazole single crystal. *The Journal of Chemical Physics*, 52(6):3121, 1970.

[76] S.J. Paddison, K.D. Kreuer, and J. Maier. About the choice of the protogenic group in polymer electrolyte membranes: Ab initio modelling of sulfonic acid, phosphonic acid, and imidazole functionalized alkanes. *Physical Chemistry Chemical Physics*, 8(39):4530–4542, 2006.

[77] M. Vasconcelos and A.A.S.C. Machado. Simultaneous determination of the acid and basic ionization constants of imidazole. *Talanta*, 33(11):919–922, 1986.

[78] I. Ivanov and M.L. Klein. Deprotonation of a histidine residue in aqueous solution using constrained ab initio molecular dynamics. *Journal of the American Chemical Society*, 124(45):13380–13381, 2002.

[79] I. Ivanov, B. Chen, S. Raugei, and M.L. Klein. Relative pKa Values from First-Principles Molecular Dynamics: The Case of Histidine Deprotonation. *Journal of Physical Chemistry B*, 110(12):6365–6371, 2006.

[80] C.M. Maupin, K.F. Wong, A.V. Soudackov, S. Kim, and G.A. Voth. A Multistate Empirical Valence Bond Description of Protonatable Amino Acids. *Journal of Physical Chemistry A*, 110(2):631–639, 2006.

[81] A.P. Kirilova, V.D. Maiorov, A.I. Serebryanskaya, N.B. Librovich, and E.N. Guryanova. Ion-molecular composition of the methane sulfonic acid-water system from IR spectroscopic data. *Russian Chemical Bulletin*, 34(7):1366–1371, 1985.

[82] A.K. Covington and R. Thompson. Ionization of moderately strong acids in aqueous solution. Part III. Methane, ethane, and propane sulfonic acids at 25 C. *Journal of Solution Chemistry*, 3(8):603–617, 1974.

[83] A. Telfah, G. Majer, K.D. Kreuer, M. Schuster, and J. Maier. Formation and mobility of protonic charge carriers in methyl sulfonic acid-water mixtures: A model for sulfonic acid based ionomers at low degree of hydration. *Solid State Ionics*, 181(11-12):461–465, 2010.

[84] L. Wang. Clusters of Hydrated Methane Sulfonic Acid $CH_3SO_3H \cdot (H_2O)$ n (n= 1- 5): A Theoretical Study. *Journal of Physical Chemistry A*, 111(18):3642–3651, 2007.

[85] S. Li, W. Qian, and F.M. Tao. Ionic dissociation of methane sulfonic acid in small water clusters. *Chemical Physics Letters*, 438(4-6):190–195, 2007.

[86] S.J. Paddison. Proton Conduction Mechanisms at Low Degrees of Hydration in Sulfonic Acid-Based Polymer Electrolyte Membranes. *Annual Review of Materials Research*, 33(1):289–319, 2003.

[87] V.A. Glezakou, M. Dupuis, and C.J. Mundy. Acid/base equilibria in clusters and their role in proton exchange membranes: computational insight. *Physical Chemistry Chemical Physics*, 9(43):5752–5760, 2007.

[88] Y.K. Choe, E. Tsuchida, T. Ikeshoji, S. Yamakawa, and S. Hyodo. Nature of proton dynamics in a polymer electrolyte membrane, nafion: a first-principles molecular dynamics study. *Physical Chemistry Chemical Physics*, 11(20):3892–3899, 2009.

[89] J.C. Perrin, S. Lyonnard, and F. Volino. Quasielastic Neutron Scattering Study of Water Dynamics in Hydrated Nafion Membranes. *Journal of Physical Chemistry C*, 111(8):3393–3404, 2007.

[90] T.A. Zawodzinski, J. Davey, J. Valerio, and S. Gottesfeld. The water content dependence of electro-osmotic drag in proton-conducting polymer electrolytes. *Electrochimica Acta*, 40(3):297–302, 1995.

[91] K.D. Kreuer. On the development of proton conducting materials for technological applications. *Solid State Ionics*, 97(1-4):1–15, 1997.

[92] R. Devanathan, A. Venkatnathan, and M. Dupuis. Atomistic simulation of nafion membrane: I. Effect of hydration on membrane nanostructure. *Journal of Physical Chemistry B*, 111(28):8069–8079, 2007.

[93] R. Devanathan, A. Venkatnathan, and M. Dupuis. Atomistic simulation of Nafion membrane. 2. Dynamics of water molecules and hydronium ions. *Journal of Physical Chemistry B*, 111(45):13006–13013, 2007.

[94] S.S. Jang, V. Molinero, T. Çaın, and W.A. Goddard III. Nanophase-segregation and transport in Nafion 117 from molecular dynamics simulations: effect of monomeric sequence. *Journal of Physical Chemistry B*, 108(10):3149–3157, 2004.

[95] D. Seeliger, C. Hartnig, and E. Spohr. Aqueous pore structure and proton dynamics in solvated Nafion membranes. *Electrochimica Acta*, 50(21):4234–4240, 2005.

[96] S.M.J Zaidi and MI Ahmad. Novel SPEEK/heteropolyacids loaded MCM-41 composite membranes for fuel cell applications. *Journal of Membrane Science*, 279(1-2):548–557, 2006.

[97] C.S. Karthikeyan, S.P. Nunes, L. Prado, ML Ponce, H. Silva, B. Ruffmann, and K. Schulte. Polymer nanocomposite membranes for DMFC application. *Journal of Membrane Science*, 254(1-2):139–146, 2005.

[98] M.I. Ahmad, S.M. Zaidi, and S. Ahmed. Proton conducting composites of heteropolyacids loaded onto MCM-41. *Journal of Power Sources*, 157(1):35–44, 2006.

[99] C.T. Kresge, M.E. Leonowicz, W.J. Roth, J.C. Vartuli, and J.S. Beck. Ordered mesoporous molecular sieves synthesized by a liquid-crystal template mechanism. *Nature*, 359(6397):710–712, 1992.

[100] J.S. Beck, J.C. Vartuli, W.J. Roth, M.E. Leonowicz, C.T. Kresge, K.D. Schmitt, C.T.W. Chu, D.H. Olson, and E.W. Sheppard. A new family of mesoporous molecular sieves prepared with liquid crystal templates. *Journal of the American Chemical Society*, 114(27):10834–10843, 1992.

[101] R. Marschall, I. Bannat, J. Caro, and M. Wark. Proton conductivity of sulfonic acid functionalised mesoporous materials. *Microporous and Mesoporous Materials*, 99(1-2):190–196, 2007.

[102] P. Ugliengo, M. Sodupe, F. Musso, IJ Bush, R. Orlando, and R. Dovesi. Realistic Models of Hydroxylated Amorphous Silica Surfaces and MCM-41 Mesoporous Material Simulated by Large-scale Periodic B3LYP Calculations. *Advanced Materials*, 20(23):4579–4583, 2008.

[103] R. Marschall, J. Rathousky, and M. Wark. Ordered Functionalized Silica Materials with High Proton Conductivity. *Chemistry of Materials*, 19(26):6401–6407, 2007.

[104] R. Marschall, I. Bannat, A. Feldhoff, L. Wang, G.Q.M. Lu, and M. Wark. Nanoparticles of Mesoporous SO_3 H-Functionalized Si-MCM-41 with Superior Proton Conductivity. *Small*, 5(7):854–859, 2009.

[105] R. Marschall, M. Sharifi, and M. Wark. Proton conductivity of imidazole functionalized ordered mesoporous silica: Influence of type of anchorage, chain length and humidity. *Microporous and Mesoporous Materials*, 123(1-3):21–29, 2009.

[106] G. Alberti and M. Casciola. Composite Membranes for Medium-Temperature PEM Fuel Cells. *Annual review of materials research*, 33(1):129–154, 2003.

[107] M. Wark, M. Sharifi, R. Marschall, M. Wilkening, P. Tölle, C. Köhler, T. Frauenheim, and D. Wallacher. Proton Conductivity of Aluminium or Sulfonic Acid Functionalized Ordered Mesoporous Silica Materials. 12.

[108] M. Wilhelm, M. Jeske, R. Marschall, W.L. Cavalcanti, P. Tölle, C. Köhler, D. Koch, T. Frauenheim, G. Grathwohl, and J. Caro. New proton conducting hybrid membranes for HT-PEMFC systems based on polysiloxanes and SO3H-functionalized mesoporous Si-MCM-41 particles. *Journal of Membrane Science*, 316(1-2):164–175, 2008.

[109] D. Gomes, R. Marschall, S.P. Nunes, and M. Wark. Development of polyoxadiazole nanocomposites for high temperature polymer electrolyte membrane fuel cells. *Journal of Membrane Science*, 322(2):406–415, 2008.

[110] Cambridge Cluster Database. *http://www-wales.ch.cam.ac.uk/CCD.html*.

[111] S. Maheshwary, N. Patel, N. Sathyamurthy, A.D. Kulkarni, and S.R. Gadre. Structure and Stability of Water Clusters (H2O) n, n= 8- 20: An Ab Initio Investigation. *Journal of Physical Chemistry A*, 105(46):10525–10537, 2001.

[112] M.P. Hodges and A.J. Stone. Modeling small hydronium–water clusters. *The Journal of Chemical Physics*, 110(14):6766, 1999.

[113] R.E. Kozack and P.C. Jordan. Polarizability effects in a four-charge model for water. *The Journal of Chemical Physics*, 96(4):3120, 1992.

[114] E.J. Bylaska, W.A. de Jong, N. Govind, K. Kowalski, T.P. Straatsma, M. Valiev, D. Wang, E. Apra, T.L. Windus, J. Hammond, and et al. NWCHEM, a computational chemistry package for parallel computers, Version 5.1. *Pacific Northwest National Laboratory,Richland, Washington*, pages 99352–0999, 2007.

[115] Density Functional Tight Binding manual - version 1.0.

[116] C.M. Maupin, B. Aradi, and G.A. Voth. The Self-Consistent Charge Density Functional Tight Binding Method Applied to Liquid Water and the Hydrated Excess Proton: Benchmark Simulations. *Journal of Physical Chemistry B*, 114(20):6922–6931, 2010.

[117] H. Zhou, E. Tajkhorshid, T. Frauenheim, S. Suhai, and M. Elstner. Performance of the AM1, PM3, and SCC-DFTB methods in the study of conjugated Schiff base molecules. *Chemical Physics*, 277(2):91–103, 2002.

[118] A. Einstein. On the movement of small particles suspended in stationary liquids required by the molecular-kinetic theory of heat. *Annalen der Physik*, 17:549–560, 1905.

[119] I.S. Chuang and G.E. Maciel. A detailed model of local structure and silanol hydrogen bonding of silica gel surfaces. *Journal of Physical Chemistry B*, 101(16):3052–3064, 1997.

[120] J.M. Stallons and E. Iglesia. Simulations of the structure and properties of amorphous silica surfaces. *Chemical Engineering Science*, 56(14):4205–4216, 2001.

[121] W.L. Jorgensen, J. Chandrasekhar, J.D. Madura, R.W. Impey, and M.L. Klein. Comparison of simple potential functions for simulating liquid water. *The Journal of Chemical Physics*, 79(2):926, 1983.

[122] P.E.M. Lopes, V. Murashov, M. Tazi, E. Demchuk, and A.D. MacKerell Jr. Development of an Empirical Force Field for Silica. Application to the Quartz- Water Interface. *Journal of Physical Chemistry B*, 110(6):2782–2792, 2006.

[123] W. Humphrey, A. Dalke, and K. Schulten. Vmd - Visual Molecular Dynamics. *Journal of Molecular Graphics*, 14(1):33–38, 1996.

Colophon: Most figures were prepared with the program xfig and gnuplot. The program VMD[123] was used for Fig. 4.1, Fig. 4.4 and Fig. 4.5.

List of widly used Symbols in Formulae:

Symbols and General Functions

$\sum_i$	- sum over i
$\sum_i'$	- sum over i, with $i \neq 0$
ϵ_{abc}	- Levi-Cita symbol - screw symmetric tensor
δ	- Dirac delta function
δ_{ij}	- Kronecker delta
∇_i	- partial derivative
N	- number of atoms or electrons; number of atoms
M	- number of nuclei

Coordinates and Vectors

t	- time coordinate
x_a^A	- 3 dimensional space coordinate of atom A; $a \in [1,2,3]$; also $\vec{x}^A$
v_a^A	- 3 dimensional velocity of atom A; $a \in [1,2,3]$; also $\vec{v}^A$
F_a^A	- 3 dimensional force of atom A; $a \in [1,2,3]$
y_i	- 3N dimensional momentum coordinate; $i \in [1,2,3,...3N]$
r_i	- space coordinates 3N dimensional
m^A	- mass of atom A
m_i	- mass of atom belonging to coordinate i
q^A	- charge of atom A

Quantum Mechenics and Force Fields

ϱ	- electron density
ψ	- wave function
H	- Hamiltonian
$\{\psi_i\}_N$	- N electronic wave function
$\hat{O}$	- operator
i	- electron, state ...
$\mu, \nu ...$	- orbital
k^{text}	- different constants
E_{text}	- different terms of potential energy
$r_{(AB)}$	- bondlength between atom A and atom B
$\vartheta_{(ABC)}$	- angle between atom A, atom B and atom C
$\varphi_{(ABCD)}$	- dehidral between atom A, atom B, atom C and atom D

Thermodynamics and Statistics

k_B	- Boltzmann constant
T	- Temperature
V	- Volume
P	- pressure
E	- potential Energy
P_{ab}	- pressure tensor (3 dimensional squared matrix)
E_{ab}	- energy tensor (3 dimensional squared matrix)
Σ_{ab}	- virial tensor (3 dimensional squared matrix)
Φ	- (interaction) potential
Z	- canonical partition function
Ω	- density of states
ρ	- probability distribution of states
ζ	- reaction coordinate

List of all Abbreviation

BE	-	binding energy, see Sec. 3.1 (page 43)
cc-pVDZ	-	name of DFT Basis set, see Sec. 1.1.2 (page 13)
cc-pVTZ	-	name of DFT Basis set, see Sec. 1.1.2 (page 13)
cc-pVQZ	-	name of DFT Basis set, see Sec. 1.1.2 (page 13)
DFT	-	density functional theory, see Sec. 1.1.2 (page 12)
DFTB	-	density functional based tight binding (method), see Sec. 1.1.2 (page 13)
FF	-	force field (description), see Sec. 1.1.1 (page 8)
L	-	link atom, see Sec. 1.1.2 (page 17)
LCAO	-	linear combination of atomic orbital approach, see Sec. 1.1.2 (page 11)
LDA	-	local density approximation, see Sec. 1.1.2 (page 13)
M	-	MM atom, see Sec. 1.1.2 (page 17)
mCEM	-	modified center of excess charge coordinate, see Sec. 1.4 (page 27)
MD	-	atomistic molecular dynamics (simulation), see Sec. 1.2 (page 18)
mio-1-o	-	name of a SK-file set, see Sec. 3.1.1 (page 41)
MP2	-	Møller-Plesset perturbation theory, see Sec. 1.1.2 (page 12)
MM	-	molecular mechanical (coupling), see Sec. 1.1.2 (page 17)
MSD	-	mean square displacement, see Sec. 3.2.2 (page 52)
NPT	-	isothermal-isobaric ensemble, see Sec. 1.2.2 (page 20).
NVE	-	micro canonical ensemble, see Sec. 1.2.2 (page 20).
NVT	-	canonical ensemble, see Sec. 1.2.2 (page 20).
OPLS	-	optimized potentials for liquid systems force field, see Sec. 1.1.1 (page 8)
PA	-	proton affinity, see Sec. 3.1 (page 43)
pbc	-	periodic boundary conditions, see Sec. 1.1.1 (page 8) and Sec. 1.1.3 (page 20)
PEM	-	polymer electrolyte membrane, see introduction (page 1)
PME	-	particle mesh Ewald (approach), see Sec. 1.1.1 (page 8)
Q	-	QM atom, see Sec. 1.1.2 (page 17)
QM	-	quantum mechanical molecular mechanical (coupling), see Sec. 1.1.2 (page 17)
QM/MM	-	quantum mechanical molecular mechanical (coupling), see Sec. 1.1.2 (page 17)
RDF	-	radial distribution function, see Sec. 3.2.3 (page 52)
SCF	-	self consistent field calculation, see Sec. 1.1.2 (page 11)
SK-file	-	Slater-Koster file, i.e. DFTB parameters, see Sec. 1.1.2 (page 13)
TIP3P	-	water model, see Sec. 4.1.2 (page 62)
HF	-	Hartree-Fock (method), see Sec. 1.1.2 (page 10)
GGA	-	generalized gradient approximation, see Sec. 1.1.2 (page 13)
B3LYP	-	Becke-three-parameter-Lee-Yang-Parr functional, i.e. name of DFT functional, see Sec. 1.1.2 (page 13)
6-31g*	-	name of DFT Basis set, see Sec. 1.1.2 (page 13)
pcb-1-0	-	name of a SK-file set, see Sec. 3.1.1 (page 41)
WHAM	-	weighted histogram analysis method, see Sec. 1.3.3 (page 26)

Die VDM Verlagsservicegesellschaft sucht für wissenschaftliche Verlage abgeschlossene und herausragende

Dissertationen, Habilitationen, Diplomarbeiten, Master Theses, Magisterarbeiten usw.

für die kostenlose Publikation als Fachbuch.

Sie verfügen über eine Arbeit, die hohen inhaltlichen und formalen Ansprüchen genügt, und haben Interesse an einer honorarvergüteten Publikation?

Dann senden Sie bitte erste Informationen über sich und Ihre Arbeit per Email an *info@vdm-vsg.de*.

Sie erhalten kurzfristig unser Feedback!

VDM Verlagsservicegesellschaft mbH
Dudweiler Landstr. 99
D - 66123 Saarbrücken

Telefon +49 681 3720 174
Fax   +49 681 3720 1749

www.vdm-vsg.de

Die VDM Verlagsservicegesellschaft mbH vertritt

Printed by Books on Demand GmbH, Norderstedt / Germany